COLOUR LIBRARY OF SCIENCE

BEGINNINGS OF LIFE

COLOUR LIBRARY OF SCIENCE

BEGINNINGS OF LIFE

ROBERT & MAURICE BURTON

ORBIS · LONDON

First published in Great Britain 1986 by
Orbis Book Publishing Corporation Ltd.
A BPCC plc company

Printed in Italy
10 9 8 7 6 5 4 3 2 1

**British Library Cataloguing in Publication
Data**

Burton, Maurice
    The beginnings of life.—(Colour library
    of science)
    1. Evolution
    I. Title   II. Burton, Robert, *1941–*
    III. Series
    575   QH366.2
    ISBN 0-85613-933-5

Previous pages
Two sea urchins
spawning. They have
come close together so
that the eggs and
sperm will be released
close to each other, so
increasing the chances
of fertilization.

Editor Penny Clarke
Designer Roger Kohn

Note There are are some unusual words in this book. They are explained in the Glossary on pages 62–63. The first time a word is used in the text it is printed in *italics*.

◄The female European cuckoo builds no nest. Instead she lays her eggs in the nests of smaller birds such as dunnocks and flycatchers, who then hatch and rear the young cuckoo in place of their own young.

WHAT IS REPRODUCTION?

REPRODUCTION IS ...

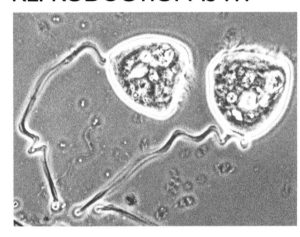

►Although vorticella is a ciliate (some of the tiny cilia are just visible), it also has a long stalk-like flagellum.

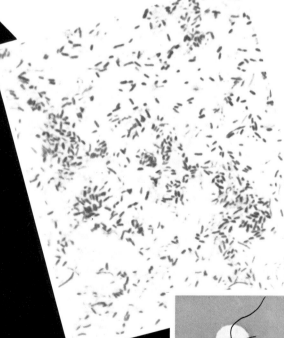

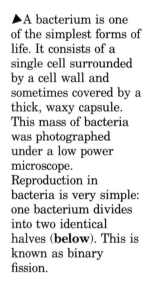

►A bacterium is one of the simplest forms of life. It consists of a single cell surrounded by a cell wall and sometimes covered by a thick, waxy capsule. This mass of bacteria was photographed under a low power microscope. Reproduction in bacteria is very simple: one bacterium divides into two identical halves (**below**). This is known as binary fission.

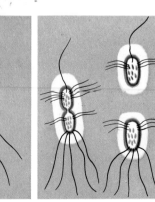

One of the definitions of living things is that they can reproduce, or make copies of themselves. A living thing which reproduces is called a parent. It produces new individuals called offspring. Adult, parent cats have offspring (kittens) which grow up into adult cats which, in their turn, have kittens. Cabbage plants make seeds which grow into new cabbage plants. Non-living things, such as rocks, are unable to do this. Reproduction is essential for living things because they have short lives. They wear out naturally and die of old age or accident, or because they are eaten by something else. As a result they have to be replaced.

The first kinds of life, scientists believe, were chemicals which could make copies of themselves. Over the course of thousands of millions of years, these chemicals became arranged in complex structures called cells that were the first living things. How this happened is very much guesswork, but single-celled living things, called *protozoans* still exist. The potozoans give us an idea of what the very earliest living things were like. Moreover, because all multicellular plants and animals are made up of many individual cells, scientists believe the protozoans help to show how single cells can behave like building blocks and so form more complicated multicellular organisms.

When a cell divides, all its different parts have to be copied. It makes no difference whether it is a single-celled organism dividing for reproduction or whether the dividing cells are in plant or animal tissues that are growing.

Reproduction in which an individual simply divides in two is called asexual reproduction. Sexual reproduction is more complicated and involves two individuals coming together. They both contribute to the production of offspring.

DIFFERENT FORMS OF REPRODUCTION

▼To the ancient Greeks, the hydra was a monster with nine heads. The modern *Hydra* is a small green or brown freshwater animal related to sea anemones. It grows asexual buds at various points on its body. Each bud breaks off and swims free to become a new hydra. It can also form new animals sexually, from ova and sperm.

Asexual reproduction

Single-celled organisms or protozoans, such as the amoeba, can be found in pond water and examined under a microscope. An amoeba is little more than a blob of *protoplasm* inside a thin *membrane*. It has no sense organs or nerves but all its functions are controlled by a *nucleus* in the protoplasm. When an amoeba grows to a certain size, it simply splits in two. First the nucleus divides in two and the two halves pull apart. Then the cell begins to split down the middle until it separates into two new amoebas, each with a nucleus. This sort of asexual reproduction, the splitting of one cell to form two, is called *binary fission*.

Multicelluar animals sometimes reproduce by binary fission. Sea anemones simply split in half and occasionally a starfish tears itself in two – three arms walking one way and the other two going in the opposite direction. After pulling apart, each half grows new arms to make a complete starfish.

Budding is another form of asexual reproduction. The organism grows a bud on its body which develops into a new individual. When this is fully grown, it breaks off and starts an indcpcndent life. Budding is a common form of reproduction in plants, bulbs, for example (page 25). Budding also takes place in simple animals. Coral reefs are made of the lime skeletons laid down by millions of sea anemone-like animals. These reproduce by budding and a single large lump or head of coral begins with one animal which divides repeatedly until a colony of thousands has formed.

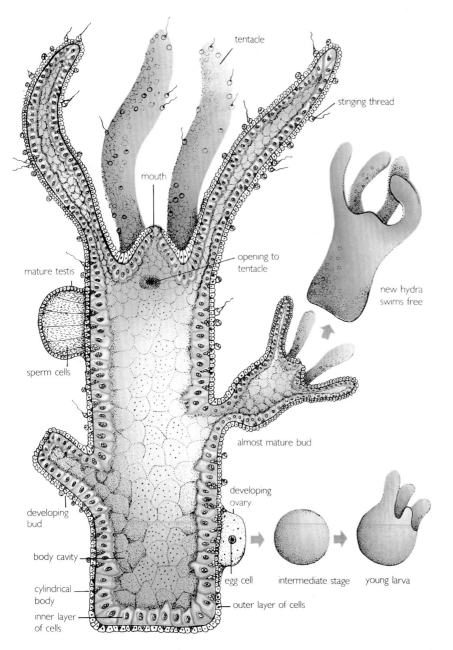

mature testis

sperm cells

developing bud

body cavity

cylindrical body

inner layer of cells

tentacle

mouth

stinging thread

opening to tentacle

new hydra swims free

almost mature bud

developing ovary

egg cell

intermediate stage

young larva

outer layer of cells

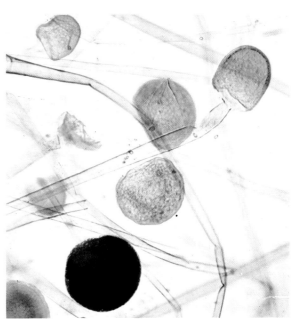

▲A fungus, as it appears under the microscope, consists of a mass of fine threads called hyphae (singular hypha). From this arise sporangia or spore-bearing fruits. These may be small, even microscopic, like these blue-green sporangia, or large, as in mushrooms and toadstools. Each sporangium produces spores, which are asexual reproductive bodies.

►A sea fan, one of the few types of coral that grow in cool seas. Like all corals, the sea fan builds up its large colony by budding, a form of asexual reproduction.

▲If a starfish loses an arm, or ray, another will usually grow and replace it. The small growth in this photograph is the new ray beginning to grow. Occasionally starfish reproduce by splitting completely in two. Each part then grows several new rays.

►The very simplest forms of life, whether plant or animal, reproduce asexually. This long thin algae, on which a rotifer is feeding, is made of a long chain of single cells.

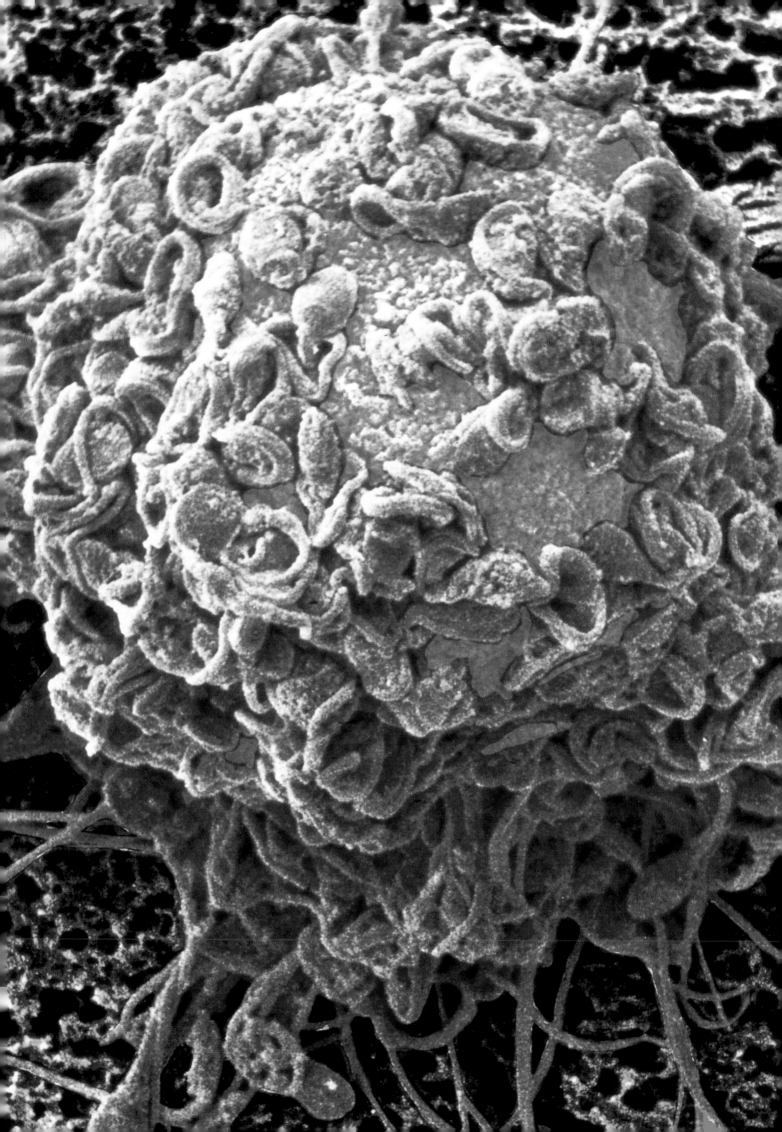

Sexual reproduction

In sexual reproduction, two cells, called *gametes*, fuse together to form a single cell, called the *zygote*, from which the new individual grows. The simplest form of sexual reproduction is shown by some protozoans, which live in the same pond water as amoebas. Although they usually reproduce by binary fission, two individuals sometimes meet and join together. Their nuclei fuse together and a zygote is formed. This process is called fertilization. Later, the zygote divides to form new individuals. In some types of protozoan, some cells are smaller than others. The small ones are male cells, and are called spermatozoa, or sperm. The larger ones are female cells, and are called ova (singular: ovum) or eggs. Sperm and ova are sex cells or gametes.

All large, multicellular organisms produce sperm and eggs in special parts of the body. Sperm are made in the testis and ova in the ovary. The ova do not move and the sperm must swim to them, either through water or in a special fluid. Each sperm has a *flagellum*, which looks like a tiny lashing whip. The lashing of the flagellum propels the sperm through the liquid to the ovum, although what guides them in the right direction is still unknown.

A mass of sperm gather around the ovum but only one penetrates it. The head of the sperm becomes attached to the surface of the egg, which dissolves to allow the head of the sperm to enter. Once the sperm is inside, the ovum forms a solid membrane which prevents other sperm from entering. The nucleus of the sperm then fuses with the nucleus of the ovum, and the zygote is formed. From this single cell, which is a combination of both parents, the new individual grows. The developing individual is called an *embryo*. The process of development of an embryo is different in plants and in animals.

◄The human ovum is no larger than the full stop at the end of this sentence. This micrograph (photo taken through a microscope) shows an ovum (pink) surrounded by the much smaller sperm (brown). Egg and sperm have been stained with dyes to make them show up more clearly.

Life cycles

All plants and animals have a life cycle. This is the series of stages through which every individual passes in its lifetime. Unless they are the product of asexual reproduction, all animals and plants start as a fertilized ovum. They develop as an embryo until the plant seed germinates, the egg hatches or the young are born. The young organism then goes through a period of development and growth, until it becomes a sexually mature adult. In some animals the young are very different in appearance from the adult and are known as *larvae* (singular: larva). For instance, a tadpole is the larva of a frog and a caterpillar is the larva of a butterfly or moth. The adult individual will reproduce one or more times in its lifetime. Annual plants, for instance, flower and set (produce) seed only once before they die, whereas trees produce seeds, known as fruits, every year for many years.

Hermaphrodites

Most animals are either male or female, but in some species each individual has both male and female sex organs. Such an animal is called an hermaphrodite. Most hermaphrodites do form pairs to mate, but they exchange sperm so that each one fertilizes the other's eggs. Earthworms, snails and slugs are examples of hermaphrodites that do this. Occasionally the hermaphrodite can fertilize itself. Tapeworms live in the intestines of other animals. They must fertilize themselves because it may be impossible for one tapeworm to meet another.

▼Earthworms are hermaphrodite. That is, each is male and female, producing eggs and sperm. But an earthworm cannot fertilize its own eggs. It must mate with another earthworm. The two emerge from their burrows and reach out across the surface of the ground to lie side by side. Each gives out sperm that fertilize the eggs of the other.

11

◀The greenfly or aphid in the centre of the picture is giving birth to a young one without having mated. This type of reproduction is known as parthenogenesis.

▼Two chains of slipper limpet forming on a stone. These limpets start life as males, but when another slipper limpet settles on them, they become female.

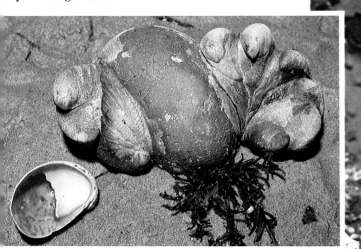

Parthenogenesis

Asexual reproduction is very rare among larger, more advanced animals but a few reproduce by a special form of asexual reproduction known as parthenogenesis or virgin birth. In these animals, the eggs develop without being fertilized by sperm and grow into adults in the normal way. Parthenogenesis is found in some species of water fleas, insects, worms and molluscs. The best known example of parthenogenesis is the aphid.

Aphids are bugs that suck the sap of beans, roses and other plants. When one aphid lands on a bean plant during the summer, it starts giving birth to young aphids, as many as 25 in one day. Because the eggs have not been fertilized, they are all females. These daughters soon start reproducing and the plant is quickly covered with aphids.

Parthenogenesis is, therefore, a way in which a single animal can give rise to a whole new population. At the end of the summer, the aphids start to produce males as well as females. Mating now takes place and the females lay fertilized eggs which hatch in the following spring.

Parthenogenesis also occurs in a few fish, salamanders and lizards. There are usually a few males, but most reproduction is accomplished without them.

Sex changes

A few animals change sex during the course of their lives. One of these is the slipper limpet that lives on seashores. This is related to the periwinkles but it looks more like a limpet. All slipper limpets start as males but when one settles on a rock, like a true limpet, it changes into a female. Then another settles on the first and remains male. It fertilizes the female under it. A third slipper limpet settles on the second and so on until a stack of shells forms. Each one starts as a male and becomes female when another settles on it.

Several fish undergo sex changes. The cleaner wrasse which lives on coral reefs is one example. Cleaner wrasse live in small groups of eight to ten fish. All these are females except one, which is a male. If the male dies, one of the females changes sex. Within a few hours of the male's death she is behaving like a male and in a few days is capable of fertilizing the remaining females.

▲A cleaner wrasse working in the gills of a squirrel fish. If the male leader of a group of these wrasse dies, one of the females will become male.

2 REPRODUCTION IN THE PLANT WORLD

SIMPLE FORMS OF REPRODUCTION

The simplest plants are not easy to distinguish from the simplest animals. Both are single-celled organisms, only a fraction of a millimetre long, which can be found living in fresh water. But those which are classed as animal, such as amoeba, have to catch and eat their food, as large animals do, while the plant-like single-celled organisms manufacture food by *photosynthesis*. This is the process by which all green plants use sunlight to turn the gas carbon dioxide from the air and water into *carbohydrates*.

One of the commonest simple plants, the euglena, spins through the water, propelled by the beating of a whip-like flagellum. Like the amoeba, the euglena reproduces only by binary fission. It rests on the surface of the water, sometimes in large numbers, and secretes a covering of slime. Each euglena then splits down the middle and the two halves form new individuals.

The simplest multicellular plants are little more than a group of single-celled organisms joined together to form a colony. Volvox is a good example. It is a hollow ball of up to 20,000 cells, though the whole colony is only as big as a pin-head. When reproducing sexually, some cells in the colony divide repeatedly to form male gametes or sperm. These swim to another volvox within the colony where they fertilize cells which have become female gametes or egg cells. The zygote eventually breaks out of the parent and grows into a new volvox. Volvox can also reproduce asexually by the division of certain cells to form 'daughter' organisms.

ALGAE

Reproduction is more complicated in the algae. These are plants which usually live in water, although a few live in damp places on land. The brown seaweeds that cover the rocks on the seashore are algae. One of the commonest is the bladder wrack whose fronds have air-filled bladders so that they float at high tide. The tips of the fronds divide into two and some grow a cushion-like receptacle in which the sex organs develop. The receptacles are orange on male plants and red on female ones. The sex organs develop in pits called conceptacles. At low tide the gametes are squeezed out, so that when the tide returns the sperms can swim in search of the egg cells and fertilize them.

▶Algae may consist of a single cell, like the chlamydomonas. Others, like volvox, are made up of a colony of single cells. Seaweeds are also algae, and the largest of them may be over 30 m (90 ft) long. Although seaweeds are more complicated than chlamydomonas and volvox their structure is still simpler than that of ferns or flowering plants.

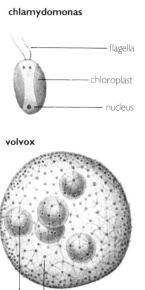

chlamydomonas

— flagella

— chloroplast

— nucleus

volvox

— ordinary cell
— daughter cell

▶The air-filled bladders that give the seaweed bladderwrack its name are clearly visible in this photograph. Seaweeds are a type of algae.

◻A large rock pool covered with another form of seaweed. The stems of this one are hollow to help it float at high tide. Like the bladderwrack **above**, this seaweed is an alga.

TWO-PART LIFE CYCLES

The simplest land plants are mosses and liverworts which usually live in damp places. Yet even those which live in dry places can reproduce only when it is wet because the male gametes need water in which to swim to the female gametes. Mosses and liverworts have a two-stage life cycle in which a sexual stage is followed by an asexual stage. This is called *alternation of generations*. The moss or liverwort plants which grow on damp walls, on tree trunks or on the surface of the soil form the sexual stage or generation. Sperm develop in tiny structures on stalks and the egg cells are held in vase-shaped organs on the surface of the plant.

When the surface of the plant is wet, the sperm swim to the egg cells and fertilize them. The zygote then grows into

▼Damp woodlands such as this one are good places to find plants that have two-part life cycles. The magnificent fronds of this fern are the asexual stage of the plant's reproductive cycle.

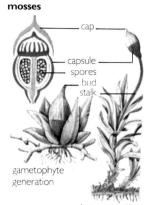

mosses

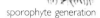

gametophyte generation

sporophyte generation

cap
capsule
spores
bud
stalk

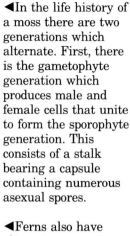

◀In the life history of a moss there are two generations which alternate. First, there is the gametophyte generation which produces male and female cells that unite to form the sporophyte generation. This consists of a stalk bearing a capsule containing numerous asexual spores.

ferns

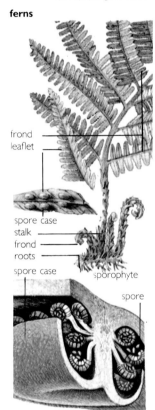

frond
leaflet

spore case
stalk
frond
roots
spore case

sporophyte

spore

◀Ferns also have alternate generations. The ferns such as those (**below left**) represent the sporophyte or spore-bearing generation. On the backs of the fronds are brown spore cases. When the spores are ripe they drop from their cases and each spore may grow into a heart-shaped prothallus. This is the gametophyte generation, which is very small and easily overlooked. On its upper surface it bears male and female cells which come together to give rise to a new fern plant.

a long stalk with a capsule on the top. This is the asexual stage. *Spores* develop within the capsule. When the spores are ripe, the capsule opens and the spores are flicked out as the capsule moves in the breeze or is knocked by animals. The wind carries the spores away. If they land in a suitable place, the spores grow into new plants.

Ferns also have a two-stage life cycle but the plants that we see are the asexual stage. The sexual stage, or *prothallus*, is very small. It grows in damp places because sperm, too, must have water to swim in to reach the egg cells. After its egg cells have been fertilized, each grows into a large plant with leaves or fronds. The spores develop in capsules on the underside of the fronds.

CONIFERS

The algae, mosses, liverworts and ferns need moisture for sexual reproduction, but the cone-bearers, or conifers, and the flowering plants do not.

The conifers are plants whose pollen and seeds develop in cones. Most conifers are trees or shrubs, with woody trunks and are usually evergreen. The cones are made up of special leaves called scales. The male and female organs lie on the inside of the scales. Male cones ripen in spring and the scales open so that the pollen can blow away on the wind. At the same time, the scales of the female cones, which are much larger, open so that the pollen can reach the *ovules*.

When fertilization has taken place, the ovule turns into a seed. During this time, the scales of the cone become hard and woody. When the seeds are ripe, the scales of the cone bend outwards and the seeds fall out. The seeds of pines have papery wings. These enable them to be carried by the wind away from the parent tree before they land and grow. If they

▲A larch in summer. The maturing green cones were fertilized in the spring. By the autumn they will be brown. Next spring they will release their seeds.

▶Like the larch, the Douglas fir is a gymnosperm. The scales on the cones of this fir have opened and released their seeds.

just fell on the ground below the parent the tiny new seedlings would have little chance of survival as they competed for water and light with the adult tree.

FLOWERING PLANTS

►Flowers of the Compositae family, for example daisies, are made up of many florets (**inset**). Mature stigmas that do not receive pollen by cross-pollination curl back and so are pollinated by their own anthers. They are said to be self-pollinated.

The flowering plants are different from the conifers because their seeds are enclosed in a container – the fruit. The flowers are the reproductive organs and are similar to the cones of conifers. Some flowering plants, primroses for example, contain both male and female organs, others are either male or female. A holly tree, for example, may be male or female, and only female trees bear berries.

A flower consists of a stem called a receptacle surrounded by the petals and sepals which are special kinds of leaves. Petals are usually brightly coloured, and in some species of flowering plant they are joined to make a tube. Daffodils are

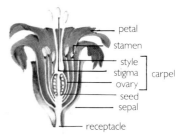

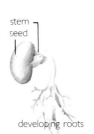

◄Flowering plants are the most complex of plants. Pollen from stamens in the flower represents the male cells. When pollen fall on the pistil the ovules are fertilized and seed is formed.

a good example. The *sepals* are usually small and green. At first, the petals and sepals form a bud to protect the rest of the flower. Then they open out to expose the *stamens* which carry the pollen grains in the *anthers* and the *carpels* which contain the ovules.

Pollination and fertilization

Fertilization takes place when pollen grains land on the *stigma*. Each grain sends out a pollen tube which grows down the *style* and enters an ovule. A nucleus in the tip of the pollen tube fuses with the nucleus in the ovule to form the zygote.

Pollen is usually carried from one flower to fertilize another – a process called *cross-pollination*. When pollen fertilizes the stigma of the same plant, the process is called *self-pollination*. In some plants, such as grasses and many trees, the grains are so small and light they can be carried by the wind. The stigma is large and feathery to trap the grains as they float past, but huge amounts of pollen have to be made because most of it is blown away and wasted.

Other plants are pollinated by insects. Less pollen is needed because it is carried from one flower to another. The flower attracts bees, butterflies, moths or flies by its brightly coloured petals, its scent and the sugary nectar which is made in *nectaries* at the base of the petals. The pollen is sticky and the insects pick it up on their bodies. When they visit another flower the pollen is brushed onto the stigma. In the tropics humming birds

◀Two bees covered with pollen collecting nectar from a thistle. In nature, chance alone determines on which plant the pollen of another lands. That is why plant breeders pollinate plants artificially when they are developing new types of plant.

▲Grass plants, are wind-pollinated. Their pollen is very light, so it is easily blown by the wind.

19

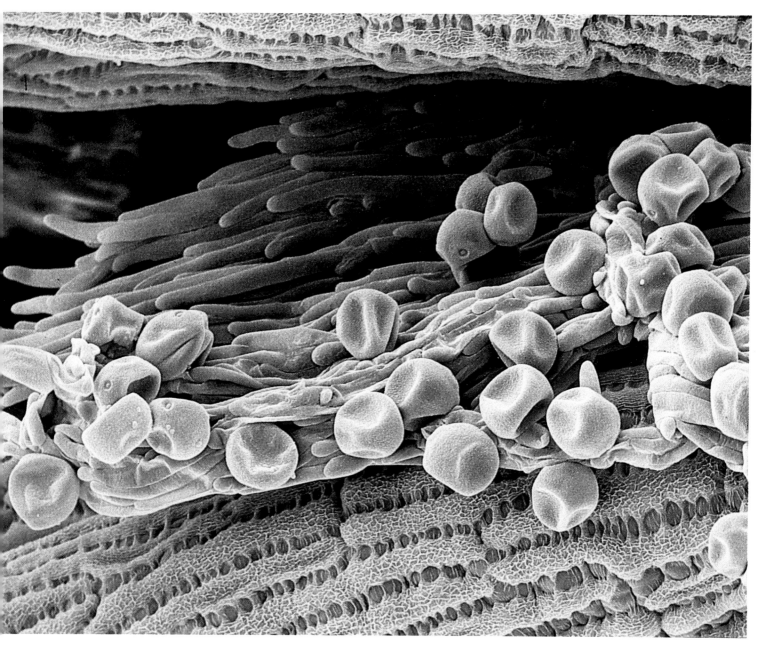

◀The slim beak of this sunbird is ideal for reaching the nectar in these narrow flowers. As the bird goes from plant to plant, it helps pollinate them.

▲Pollen grains come in many shapes and sizes. This photomicrograph shows the pollen of annual meadow grass.

There is also a store of food called the endosperm which will nourish the growing plant.

The carpel that surrounds the seeds becomes the fruit which can be soft and juicy, as in a plum, or dry, like a hazelnut or acorn.

▼Laurel berries at different stages of ripeness – from green to purple when fully ripe.

pollinate plants such as lilies and tobacco and bats pollinate sisal and the baobab. A few water plants, for example Canadian pond weed, have pollen which is carried by the water.

The embryo forms

After fertilization, the ovule turns into a seed with an embryo inside. The embryo is the young plant and it has a plumule (a small shoot), a radicle (a small root), and one or two cotyledons or seed leaves.

SPREADING THE SEEDS

▼Some of the seeds of this dandelion have already been blown away by the wind. The parachute of feathery hairs ensures that the seeds will be carried far from the parent plant.

When the seeds are ripe they are shed by the plant. Unless they land on suitable soil they will not germinate or sprout. Even then, a long time may pass until development of the new plant starts. In deserts, seeds may not sprout until it rains, perhaps years later and, in forests, seeds may not sprout until the tree above them dies.

Wind-borne seeds

Many plants have seeds which are carried long distances. Gorse and broom seeds are shot out by the twisting of the seed pods as they dry and open with a sharp crack. Some, like the foxglove and the orchids, have tiny, very light seeds which are carried in the wind, and others have 'parachutes' or 'wings' which help to keep them airborne. Dandelion seeds have a plume of fine hairs and sycamore seeds have wings. If there is a good wind they are carried a very long way. If it starts to rain, they immediately fall to the ground and are washed into the soil, where they will germinate.

▼Sycamore fruits hang in pairs, each part containing one plump seed. When the seed reaches the ground the protective case does not split. Instead it slowly decays, so the seed can germinate the next spring – when conditions for growth are best.

▲The seed pods of the false acacia or locust tree. The pods open, releasing the small black seeds, in the autumn.

▶A sycamore seedling can grow up to 50 cm (20 in) a year for its first few years. This can make them a nuisance in woods, where they will crowd out other young trees if left undisturbed.

Animal carriers

Other kinds of seed are too heavy to be blown by the wind. When they fall off the plant, they land on the ground beneath. They rely on animals to carry them. Seeds of cleavers, or goosegrass, and burr marigold have seeds with tiny hooks which catch in animals' fur. Many other kinds of seed are carried in mud that sticks to animals' feet. Humans will also spread seeds in this way and the seeds of plants that grow beside roads may be carried in the treads of vehicle wheels.

Nuts and berries are *fruits* which have flesh surrounding the seed. The flesh is dry in nuts and juicy in berries. Animals eat the flesh and the seeds either pass through the animals' bodies unharmed and so are carried away from the parent plant, or, in the case of nuts, are buried by the animals for future use. Many animals and birds feed almost entirely on fruit, for example parrots, toucans and fruit pigeons, and mammals such as fruit bats and monkeys of many kinds. The bright appearance and sweet taste of many fruits are a special attraction to these animals. The colour and taste develop when the fruit is ripe so that it is not eaten before the seeds are properly formed.

Sometimes the fruits are spread when the animals carry the fruits and hide them. Squirrels bury nuts in the ground and if they forget them, the nuts germinate and grow. Jays bury acorns, another example of the way plant seeds are spread.

Mistletoe is a plant which grows on trees. Its berries are so sticky that birds have to remove the flesh from their beaks by wiping them on the branches where they are perching. This action sticks the seeds to the bark.

Carried by water

A few seeds are carried by water. Plants such as waterlilies have floating seed pods, and alder trees which grow along riverbanks have floating seeds. Coconuts float across oceans, so that coconut palms are found on tiny remote islands thousands of kilometres from other land.

▲The dog may be uncomfortable with an ear full of burrs, but the picture shows the efficiency of this method of seed dispersal.

▼The hooks on the ends of these wood avens fruit will catch in clothing or animal fur, so spreading the plant's seeds.

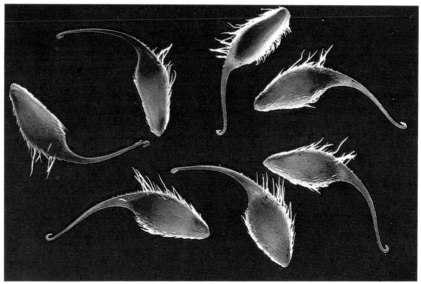

◀Acorns fall in the autumn and may germinate if the ground is damp. But no leaves will apppear until the spring. Oak seedlings grow about 15 cm (6 in) in 6 months – more slowly than the sycamore (**opposite**).

23

VEGETATIVE REPRODUCTION

▼A good crop of strawberries! The strawberry has two forms of reproduction, from the flowers come the fruit bearing many seeds (**above**). The reproduction by seeds is sexual. In addition, a strawberry plant throws out runners. These take root, so producing new plants. This is called vegetative reproduction and is an asexual form of reproduction.

▼The runners which will form new strawberry plants are clearly visible across the sheet of black polythene covering this strawberry bed.

Besides flowering and producing seeds, some plants reproduce asexually. They grow new parts which eventually separate to form new plants. This is called *vegetative reproduction*. It is a way in which a single plant grown from a seed can form a new colony of plants. Gardeners use vegetative reproduction as a method of quickly growing new plants, but many of the plants that gardeners regard as weeds spread rapidly by vegetative reproduction.

Runners, rhizomes and tubers
Plants such as strawberries and blackberries send out runners – long stems that lie along the ground. Where they touch the ground they sprout first roots and then buds which grow into leaves, so forming a new plant. Then the runner dies off and the new plant becomes independent. The suckers of poplar trees and mint plants behave rather as if they were runners, but they are, in fact, stems which grow underground and send up new shoots.

The rhizome is another kind of stem that grows along the ground. Irises grow from rhizomes. At the tip of each rhizome there is a cluster of leaves and a flower stalk. These die down each winter. Scars o previous years' leaves can be seen on the rhizome. Each year the rhizome grows two new branches so that it becomes shaped like the letter Y. As new branches are grown, the old parts wither and die, so that the branches become the new plants.

The potato is another kind of stem,

called a tuber. Like a rhizome, a tuber also acts as a store of food for the growing plants. It is also an important source of food. The 'eyes' of a potato are buds. If you cut a potato into pieces, each with an eye, and plant the pieces, each one will grow into a new plant.

Bulbs and corms

Bulbs and corms are two familiar examples of vegetative reproduction. They are planted in parks and gardens or in bowls indoors, and grow into flowers such as daffodils, tulips and crocuses. A corm, like that of the crocus, is the swollen base of the stem. It stores food which is used when the new leaves and flower grow in the spring. After the flower has died and the leaves have withered a new corm is made, so that a string of corms, one on top of the other, develops.

Bulbs look like corms but their structure is different. An onion is a bulb. Cut an onion down the middle, and you can see its structure. It is made up of rings of fleshy scale-leaves attached to a flat disc. In the middle of the bulb there is a bud from which the leaves and flower grow. Sometimes there is a small 'baby onion' inside the main one. This is the new bulb which grows in size and eventually separates from the original bulb.

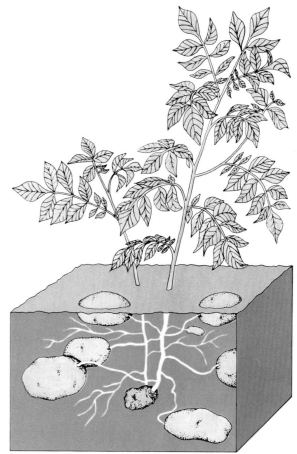

►In the diagram, the potato that the new plant grew from is the small, withered one at the base of the leaf stems. The new potatoes are growing and storing food for the next year's plants.

►A cross section of an onion showing the thickened leaves which contain the plant's store of food.

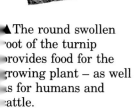

▲The round swollen root of the turnip provides food for the growing plant – as well as for humans and cattle.

►The parsnip's root is very different from the turnip's but serves exactly the same purpose – food for the new plant.

REPRODUCTION IN THE ANIMAL WORLD

Many *invertebrate* animals living in the sea reproduce by releasing their eggs and sperm into the water and leaving the sperm to swim to the eggs. This is the simplest form of sexual reproduction in the animal kingdom but it is extremely wasteful. Although the sperm cells are guided to the eggs, probably by chemicals, they cannot swim far. Moreover, many of the eggs are eaten by other animals. To make up for these losses the animals produce their eggs and sperm in huge numbers. Mussels release 25 million eggs and many more sperm, but even so some eggs never get fertilized.

Safety in numbers
To increase the chances of fertilization taking place in the enormous volume of the sea, animals such as sea urchins spawn together so that eggs and sperm in uncountable numbers are shed into the water at the same time. When one sea

SPAWNING IN THE SEA

▲Although a barnacle looks like a mollusc it is actually a crustaccan. It stands on its head and kicks food into its mouth with its legs. It sheds its eggs into the sea. There each egg grows into a nauplius larva, like the larvae of shrimps, crabs and lobsters. The nauplius swims in the sea and, after a while, grows a tough shell and changes into a cypris, again like the cypris larvae of shrimps, crabs and lobsters. Finally, it changes to an adult. It settles on a solid object on its head, grows a shell and now looks like a mollusc (**right**).

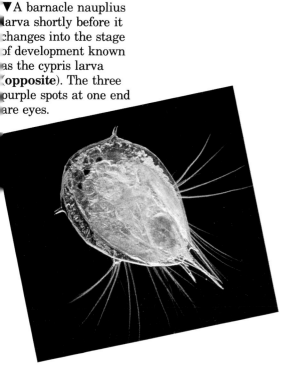

▼A barnacle nauplius larva shortly before it changes into the stage of development known as the cypris larva (**opposite**). The three purple spots at one end are eyes.

urchin spawns, a wave of spawning spreads through all the sea urchins nearby. It is triggered by chemicals secreted with the sperm of the first sea urchin to spawn. The chemicals spread through the water and into the bodies of neighbouring animals, causing them to spawn too.

Mass spawning is particularly dramatic in the bristleworm called the Palolo worm. This worm lives in burrows in coral reefs and, as spawning time draws near, the reproductive organs develop in the rear half of the body. This breaks off and swims to the surface, leaving the rest of the worm alive in its burrow. Eggs and sperm are released from the rear half at the surface which means that they are concentrated at one place, because all the worms spawn together. They spawn on successive dawns during the last quarters of the October and November moons, and the sea becomes milky with the enormous quantities of eggs and sperm.

Another bristleworm is called the fireworm. It comes to the surface of the sea to spawn 55 minutes after sunset at the third quarter of the moon and the females glow with light. The males are attracted by the light and start to flash on and off as they swim around the females. The lights of the two sexes help bring them together so that it is easier for eggs and sperm to meet.

Tiny larvae
The eggs of these and other marine creatures develop into tiny floating larvae. They have transparent, delicate bodies, on which rows of tiny hair-like projections called *cilia* beat continuously to propel the larvae as they float near the surface of the sea. Evenutally they develop into the adult animals and settle on the sea bed as limpets, mussels, barnacles, starfish, worms and many others. These animals are either slow-moving or remain fixed to one place, so their floating larvae enable them to spread and colonize new places.

INTERNAL FERTILIZATION

In the methods of fertilization described so far, wastage of eggs is very high. To reduce this the eggs of some kinds of invertebrate remain inside the female's bodies. Instead of shedding millions of tiny eggs they produce fewer, bigger eggs. Each egg now has a larger food store or yolk to feed the developing embryo. To make reproduction even less haphazard, males and females now come together and the sperm are placed inside the female's body. This is called internal fertilization.

Courtship
When male and female come together in the act of mating, they need to be sure that they belong to the same species. To be certain, the male courts the female. He signals to her, so that she can identify him. Courtship is also needed to make sure that the male and female are both in breeding condition. The female has to show that her eggs are ripe. Mating would be a waste of time if the eggs had not fully developed or if they had already been fertilized.

There are many ways in which courtship signals are exchanged. Some animals use smells, others use sounds, while others give visual signals. The fiddler crabs which live on tropical beaches signal by waving their claws. The males have one very large bright claw and the members of the different species wave it in a particular way. So when there are several species of fiddler crab on one beach, the females can choose a male of the right species with which to mate.

The courtship of squid and cuttlefish is particularly unusual. The sperm are carried in a packet called a spermatophore. The male squid courts by changing colour and when the female has accepted him, he uses a special long arm, called the hectocotylus, to place the spermatophore in her body. Male cuttlefish take on a pattern of black stripes, like a zebra, and stroke the females with their arms.

The importance of internal fertilization

Internal fertilization is essential if animals are to live on land, because otherwise the sperm would not be able to reach the eggs. All completely land-living animals, such as mammals, birds, insects, spiders and snails have internal fertilization, whereas amphibians and land crabs must return to water to breed. Some periwinkles living on the shore can mate when the tide is out because their eggs are fertilized internally and hatch into tiny periwinkles rather than larvae, but these periwinkles still need an occasional wetting by the sea to stop them drying out. Land snails and slugs have more watertight bodies than periwinkles, so they can spend their whole lives on dry land. They lay eggs with shells so the eggs, too, do not dry out, which means these animals do not need water even for reproduction.

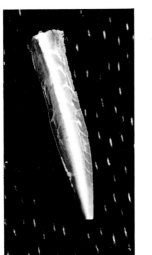

▲Garden snails are hermaphrodite. A pair will circle each other, drawing close together, then each fires a love-dart (**left**) into the other. This makes them more ready to mate. Finally, they wrap their bodies round each other and exchange sperm.

▶A garden snail lays its eggs in a hollow in the soil. The eggs look like a cluster of small pearls.

◄When insects mate, the male passes sperm into the female. The sperm is stored in a special sac in the female's body until the eggs are ready to be fertilized. Then, one sperm passes into each egg through a pore in the egg's surface.

Broadly speaking, there are two kinds of insect: those with larvae that look like the adult and those with larvae that don't.

When an insect, or any other animal, changes shape during its life, the process is called *metamorphosis*. This is from a Greek word meaning transformation. When insect larvae look like the adult insect, scientists say they have an incomplete metamorphosis because the changes are small. The second kind of insect life cycle is described as a complete metamorphosis because larva and adult are so different from each other that the larva has to undergo a complete transformation before becoming an adult.

Incomplete metamorphosis

The larvae of the first type hatch from the eggs looking very like the adult insects, except that they lack wings. These larvae grow by a series of moults. Insects cannot grow steadily because their outer skin is a tough cuticle, which will not stretch. Instead, the larvae shed their cuticles at intervals and rapidly grow in size before the new cuticle hardens. Mayflies may moult as many as thirty times.

The wings, too, grow as the larva grows. At the first moult, the larva has only the beginnings of wings, which are usually so small as to be almost

◄Glowworms are rather drab insects, but when ready to mate special organs in the female's abdomen glow very brightly to attract a male.

▲A mass of newly hatched spiderlings in the protective web of threads that the female spider spun around the eggs.

◀Termites, also known as white ants, build the largest nests of all insects. The nests, known as termitaria, form huge mounds of hard packed earth and are architectural masterpieces. The hundreds of thousands of worker termites in a nest are 'governed' by a single queen, her whitish sausage-shaped abdomen swollen with eggs, which she lays at the rate of 10,000 a day. Workers swarm over her to feed her and keep her clean. They also remove the eggs and tend them.

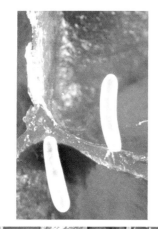

▼The queen honeybee lays her eggs (**right**) in empty cells in the honeycomb (**below**). When the eggs hatch, the young bees are fed with food stored in other cells in the comb.

▲Scarab beetles feed on dung, rolling a ball of it along to a selected spot. To the ancient Egyptians this symbolized the earth turning on its axis, so they regarded the beetles as sacred.

Arriving at the chosen spot, the beetles dig a hole to bury the dung. The female lays her eggs in the dung and when the larvae hatch they are surrounded by food.

◀The map butterfly lays its eggs in columns on the underside of nettle leaves.

▶The caterpillar of the lobster moth feeds mainly on beech leaves. The ridges along its back and the enlarged rear make the caterpillar seem larger than it is – a useful way of confusing predators.

▼Pupae of large tortoiseshell butterflies, shortly before the adults emerge in early July.

Bottom The hairy body of the garden tiger moth is clearly visible in this close-up photograph.

unnoticeable. The wings grow bigger at each moult until, when the insect is fully grown and adult, it has wings large enough for flying.

Complete metamorphosis

The larvae of the second type of insect are generally worm-like and have no signs of wings. If the larva has no legs it is called a maggot or a grub, or if it has very small legs it is called a caterpillar. It feeds and grows, moulting its tough outer skin four or five times and increasing in size each time. Finally the grub, maggot or caterpillar goes into a resting stage. It stops feeding and growing, and does not move about.

During this resting stage it is called a pupa. In beetles, moths and butterflies the pupa is sometimes called a *chrysalis*. The pupa is covered with an extra-tough cuticle within which a most remarkable transformation takes place. All the internal organs break down to form a mush of tissue cells.

After a while the cells forming the mush begin to reorganize and form new internal organs. The wings, legs and antennae take shape, new muscles grow and the sex organs develop. Eventually the cuticle splits down the back and the perfect adult insect wriggles out, leaving only the empty cuticle, known as the pupal case.

▲Although female eels live in freshwater, they migrate to the sea to breed. The eggs hatch into tiny larvae which are carried by the currents back to the coast their parents came from. By this time they have grown into transparent elvers. The male elvers stay in the estuaries and the females migrate upstream, where they stay until they are fully grown and ready to mate.

Fish are *vertebrates*, that is, they have a backbone. They live in water, in the seas and in rivers and lakes. There are over 30,000 different species of fish and, as you would expect, since there are so many, there are many different ways in which they reproduce.

The simplest way is known as spawning. At the breeding season large numbers of herrings, for example, gather together in huge dense groups called shoals. The females shed their eggs into the water. The males shed their milt, which is another name for a mass of sperm. The sperm enter the eggs and fertilize them. The fertilized eggs begin to divide, again and again, and from each collection of cells a larva is formed.

Predators at large

When eggs and milt are shed into the water, large numbers of them are eaten by other animals. In addition, many more are destroyed by natural accidents. A herring lays a mere 20,000 to 50,000 eggs but a cod may lay up to 7 million at one spawning and a few species lay up to 30 million. However, scientists have estimated that only one in every million of these eggs survives to become an adult.

▼Dogfish are a type of small shark. They lay their eggs in tough leathery cases which have long curling threads at each corner. The threads become tangled in seaweed, so anchoring the eggs. The young dogfish are about 10–16 cm (4–6¼ in) when they hatch. The empty egg cases are often washed up on beaches, where they become black and brittle.

Safety precautions

Spawning on this scale is tremendously wasteful. It is not surprising, therefore, that some species of fish use less wasteful methods. That is, they lay fewer eggs but take greater care to protect them and the young they grow into. The wastage is offset by such methods as internal fertilization, building nests and parental care which reduces the number of eggs destroyed by predators or in accidents.

With external fertilization, there is always the danger that some eggs will not meet a sperm, but this danger is reduced by the male and female coming close together in a pair so that the sperm are shed close to the eggs. In internal fertilization, the male inserts a special organ into the female's body and releases the milt. In this way the sperm in the milt reach the eggs and fertilize them. The fertilized eggs are then kept inside the female's body for a short time before they are laid. Goldfish breed in this way. In some fish, guppies for example, the female does not lay the fertilized eggs. They remain within her body and the young are born alive.

Another method of internal fertilization is known as mouth-brooding. Cichlids, a group of fish living in the lakes and rivers of Africa, lay their eggs as usual but then take them into their mouths. They are sometimes fertilized inside the parent's mouth. Even when the young fish hatch they still shelter in the mother's mouth and, if they leave it, quickly return for safety if danger threatens.

The male stickleback builds a nest of water weeds and escorts females into it to lay their eggs. The male then enters the nest and sheds his milt to fertilize the eggs. After that the male guards the nest, fanning it with his fins to provide plenty of oxygen for the developing eggs and to keep them clean. When the young hatch, he looks after them, chasing off anything that threatens them.

Many fish living in rock pools lay their eggs in empty mussel shells to give them extra protection. Other fish wrap their own bodies around a mass of eggs to protect them. The male seahorse takes this a step further and carries the eggs in a pouch on his belly.

◄Milt containing sperm is taken from a male salmon and used to fertilize eggs.

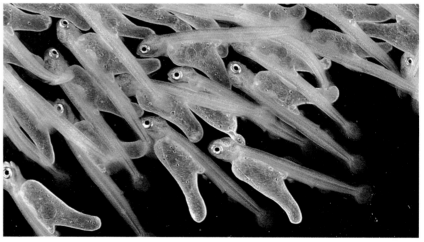

▲Young salmon, or alevins as they are called at this stage, four days after hatching. The yolk sacs, which will nourish them for the next six weeks, are clearly visible hanging below them.

AMPHIBIANS AND REPTILES

◄Four pictures illustrating the growth of a young salamander. **Top left** are several adults with eggs attached to an underwater plant. **Left** shows two embryos in their gelatinous envelopes. **Below** the young salamander is taking shape and will soon be free (**bottom**) of its protective coat.

Fish are not the only creatures that scientists describe as spawning (page 34). Amphibians also spawn. Amphibians include frogs, toads, salamanders and newts. The word 'amphibian' comes from the Greek and means 'double-life' because these animals live on land but must return to water to breed.

Related to fish
A fish looks very different from a frog because it has fins, a forked tail and breathes by means of gills. There are, however, a few fish that have leg-like fins. The mudskipper of south-east Asia is one. There are also fish that have lungs: the lungfish. And there are fish, such as the ocean sunfish, that have almost no tail at all. Scientists know from the study of fossils that amphibians evolved from fish which had leg-like limbs and lungs.

Although most amphibians shed masses of eggs or spawn into the water, some species do not. Newts, for example, wrap their eggs, one at a time, in the leaves of water plants, and some frogs and toads look after their eggs. The male midwife toad, for example, carries the spawn wrapped around his legs and returns to

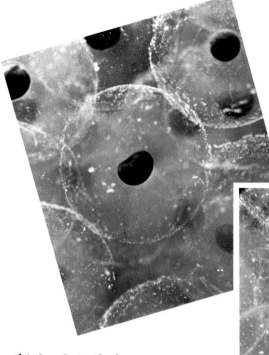

◀A mass of frog spawn (**left**) showing developing eggs, each in its transparent sphere of jelly. A few days later and the developing egg, now an embryo, is taking on the familiar shape of a tadpole (**below**).

◀A female turtle, her shell crusted with barnacles, lays her eggs in a hole she has scraped in the sand. Once the eggs are laid she returns to the sea. When the young turtles hatch (**bottom**), they are in danger from many predators and very few survive.

the water at intervals to moisten them. The dusky salamander is one of a few species of amphibians that lay eggs in damp ground. The rainfrogs of Africa have no tadpole stage — the egg develops directly into a froglet, a tiny young frog.

Reptiles

The next most advanced group of vertebrates is the reptiles. Most of these spend all their time on land. Reptiles include lizards, snakes, tortoises, turtles and crocodiles. If you compare an amphibian, such as a salamander, with a reptile, such as a lizard, there seems little difference between them. Both have long bodies and a long tail. Both have four legs with five toes on each leg, and both have lungs for breathing. But a lizard is covered with scales while a salamander is scale-less. However, the main difference between a lizard and a salamander is that the salamander must go to water to breed. The lizard lays its eggs on land or gives birth to live young, for no reptile ever has a larva. Fertilization is always internal, and the sperm are placed in the female's body in the act of mating with the aid of the male organ, or penis. This act is called copulation. All reptile eggs have a tough parchment-like shell or a hard chalky shell very like the eggs of birds.

Most reptiles lay their eggs in a hole in the ground and leave them to hatch. Turtles come out of the sea and dig nest holes in sandy beaches. Grass snakes lay

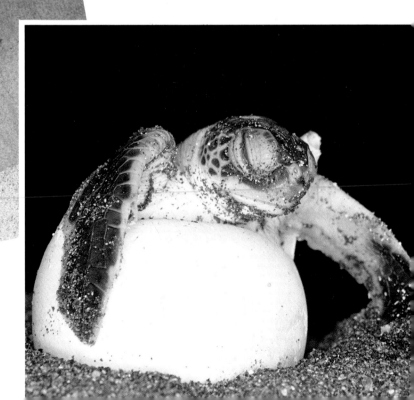

A female reticulated python guards her eggs. The reticulated python belongs to the *python* genus and is the largest of all pythons. It reaches 0 m (33 ft) in length. The young are nearly m (3 ft) long when they hatch and grow at rate of 60 cm (23 in) a ear. All female pythons take great are of their eggs. One undred or more are aid and the female raps herself round hem, leaving them nly very occasionally drink or to eat. The emperature within the ils is up to 6°C (42°F) igher than the rrounding air.

▼Young grass snakes hatching out. Reptile eggs often have rather leathery shells.

◄A bearded dragon lizard covering her eggs with sand. Immediately after mating, the female buries herself in the sand to lay her 8 to 24 eggs. The eggs do not hatch for 7 months and are not guarded by the parents, so they need the protection of being buried.

▼Three young crocodiles have hatched from this group of eggs and you can see two more of the eggs beginning to split open.

heir eggs in rotting vegetation. The heat iven off by the rotting matter keeps the ggs warm so that they develop more uickly. Pythons wrap their bodies around heir eggs to keep them warm. Keeping ggs warm is called incubation. Tortoises nd turtles are very unusual because the ex of the young depends on the emperature at which the eggs are ncubated. Crocodiles make a large nest of lants and stay to guard the eggs, lthough they do not incubate them. They lso look after their babies and carry hem to the water in their mouths.

The eggs of the European common izard and the adder hatch just before hey are laid so that the young are born live. This process is called *ovoviviparity*.

If you stop to think about it for a few moments, you'll realize that most animals reproduce by laying eggs. Only a small minority bear their young alive. Birds are the obvious example of animals that lay eggs. It is easy to see why no bird gives birth to live young: it would be difficult for the female to fly with the extra weight of babies inside.

Eggs and nests

Many birds take great care of the eggs and the young birds which hatch from them. Building a nest, as the majority of birds do, prevents the eggs from rolling about and becoming lost. A nest also gives some protection from enemies and the weather, as well as helping to keep the eggs warm. Warmth is very necessary in warm-blooded animals such as birds. If birds' eggs are allowed to get cold the chicks inside soon die.

Eggs are generally kept warm by the parent birds incubating them. The parent covers the eggs with its body which acts like a hot-water bottle. Both males and females of many birds take turns on the nest, but frequently the female does all the incubation. Incubation also gives protection from enemies and the weather.

When it is first laid, a bird's egg consists of a ball of yolk in the centre with the embryo of the chick on the surface. (This appears as a small spot on the yolk of a chicken's egg if it is fertile.) The yolk is surrounded by albumen – the 'white' of the egg – and a hard outer

◀A flamingo with its young on the nest. Flamingoes nest in colonies around the shores of lakes in warm climates. The nests are built of mud.

▼Young ostriches can run about soon after they are hatched. Birds such as ostriches are called nidifugous (page 43) by scientists.

41

►The European cuckoo is probably the best-known parasite in the bird world. **1** The cuckoo's egg beside two eggs of a Cape robin. **2** The cuckoo's egg is the first to hatch and the young bird pushes the robin's eggs out of the nest. **3** The mother robin does not realize that the young bird is an intruder, and feeds it as though it were her own. This form of behaviour means that the female cuckoo does not have to undertake the hard work of building her own nest. **4** The cuckoo is ready to fly from its foster parent's nest.

◄An adult European cuckoo perching in a hedgerow tree. Each female lays her eggs in the nest of the same two or three species of birds. The cuckoo's eggs are very similar to those of the other bird so not disturbing them in any way.

▼When the chick is ready to leave the egg it starts to tap a ring of holes in one end of the shell. It has a special egg-tooth near the tip of its beak to help it do this. The tooth is shed soon after the chick leaves the shell.

chalky shell. As the embryo develops, it uses the yolk as food. The albumen gradually disappears as the embryo gets bigger. When it is ready to hatch, the chick makes a hole in the eggshell so that it can breathe properly. It may also start to call to its mother and to the other chicks in their eggs. To hatch, the chick makes a ring of holes at the blunt end of the shell and splits it open, so that it can wriggle out. Once the eggs have hatched, the young birds are brought food by their parents until the young birds can fend for themselves.

There are two kinds of baby bird: those that can run about almost as soon as they are hatched and those that can't. The chicks of the barnyard chicken are a good example. They emerge from their eggs covered in down, their eyes open almost immediately and they soon leave the nest to start searching for food. This does not mean that the mother ignores them. She keeps them warm by brooding them under her body and she is constantly watching for danger, clucking to them all the time. Should a hawk appear overhead, she makes a special alarm call. On hearing this all the chicks run for cover.

Although the barnyard chicks are able to peck food from the time they are hatched, the hen helps them to find it.

When she finds a morsel of food, whether it is a seed or an insect, she clucks in a special way and the chicks come towards her to inspect it.

Nidifugous and nidicolous young birds

Barnyard chicks are said to be *nidifugous*, from Latin words meaning 'leaving the nest'. Most birds are *nidicolous*, which means the young stay in the nest for days or weeks after they hatch. They are naked when they hatch or at most have a few downy feathers. They are usually blind, and their eyes do not open until they are several days old. They are unable to move except to stretch their ungainly heads up on their thin scraggy necks and open wide their beaks. When the parent arrives with food, they gape so that food can be pushed down their throats.

Nidicolous babies are called nestlings because the first part of their lives is spent entirely in the nest. The parents look after them, feeding them, protecting them and keeping the nest clean by carrying away their droppings.

You may hear people say they have watched a bird teaching its young to fly. This is not really true because young birds learn the way to live largely by instinct and practice. Many birds still need help after they have learned to fly. The parent birds continue to feed and protect their young until they can look after themselves.

▼A wandering albatross with its chick in the nest. Most albatrosses nest on islands in the southern oceans. They lay one egg every two years and incubate it for 70 days.

MAMMALS

▲A pair of wild boar with their litter of young. The striped backs of the young ones help them blend with the fallen leaves on the forest floor.

▼A duckbill, or platypus, one of the two egg-laying mammals of Australia. Duckbills are water creatures and are seldom seen on land.

Birds and mammals are the only kinds of animal that can truly be called warm-blooded.

All mammals, with two exceptions, bring forth their young alive after they have developed in the mother's uterus, or womb. Another way to tell if an animal is a mammal is to see whether it has true hair on its body. Some insects seem to have hairy bodies but the hair is not like that on the bodies of mammals. On the other hand, some mammals seem

to have no hair. Elephants and rhinoceroses, for instance, have only a few scattered hairs. Whales seem to break the rule also, for their bodies are covered with a smooth, hairless skin. Yet even whales may have a few hairs on the upper lip, a kind of moustache, when they are born. All baby mammals are fed on milk which is made in the mother's body. It is produced in organs called mammary glands.

Monotremes

After reading the paragraph above you'll realize that in biology there are few firm rules that are not upset by one or two exceptions. For example, there are the two mammals that lay eggs. They are called *monotremes* and both are found in Australia. One is the duckbill or platypus, the other is the echidna or spiny anteater. But like all other mammals, and unlike birds, their babies are fed on the mother's milk.

So what is a mammal? A mammal, despite all the exceptions, is a warm-blooded animal that feeds on its mother's milk during infancy, that has true hair on its body and that, with two rare exceptions, is born live.

The platypus lives most of its time in

water. When the female is ready to lay her eggs, she digs a tunnel in the bank of a stream or lake. She never lays more than two eggs at a time. They are soft-shelled and the mother incubates them by curling herself around them, so that her body forms a sort of protective nest.

When the eggs hatch, the young duckbills seek out a spot on the underside of the mother's body near where her tail joins her body. There they find two slits in the skin from which milk oozes. They lap this, like kittens at a saucer of milk. The mother duckbill spends practically all her time sheltering with her babies in the tunnel, suckling them until they are old enough to swim and catch food, such as water insects, for themselves.

Just before laying her single egg, the female echidna develops a pouch on her abdomen. Then she lays her egg and, in a way that still puzzles scientists, she manages to get the egg into the pouch. After little more than a week the egg hatches and out comes a baby echidna. It is naked and blind but it finds its way to one of two small nipples on the upper wall of its mother's pouch and begins to suck milk. It stays in the pouch until its spines begin to grow. Then the mother leaves it in a burrow in the ground but visits it every $1\frac{1}{2}$–2 days to let it suckle.

Marsupials

There are more *marsupials* in Australia than anywhere else in the world. Marsupial is the scientific term for any kind of mammal the female of which has a pouch or marsupium on the abdomen. Kangaroos and wallabies are probably the best-known marsupials.

The babies develop inside the mother's body but they are born after only two or three weeks. At birth they are tiny and almost helpless, yet each baby marsupial can climb through its mother's fur into the pouch. There it seizes one of the nipples and begins to suck. It stays there, hanging by the nipple in its mouth for two months or so, although the young of the larger species of kangaroo may stay in the pouch for up to a year.

▼Blind and hairless, baby opossums lie in the protection of their mother's pouch. The newborn baby is unrecognizable as an opossum. Once in the pouch the tiny opossum takes one of its mother's nipples in its mouth and feeds on milk sucked from her mammary glands. The young stay in the pouch for about 10 weeks while they grow.

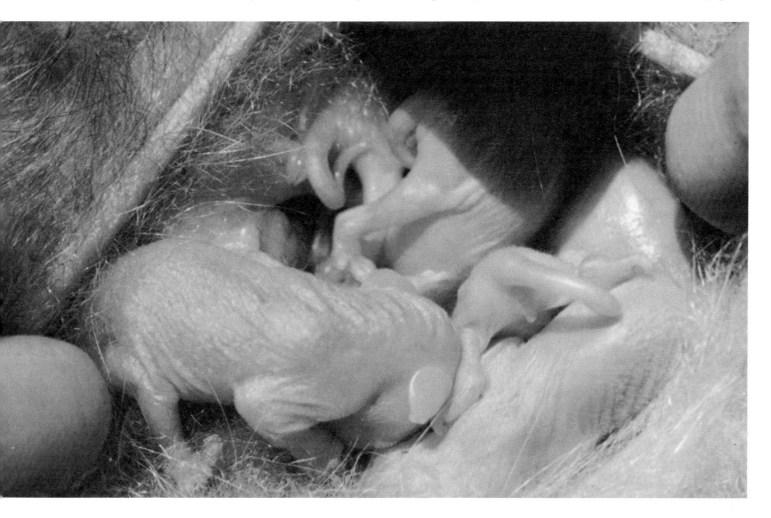

▲Even when the young kangaroo is able to jump about on its own it quickly goes back into its mother's pouch when it needs to rest. It also uses the pouch as a shelter from danger.

has been born.

The afterbirth or placenta plays a very important role in the development of young mammals. Inside it the embryo is kept warm by the heat of the mother's body. The placenta is filled with fluid so it cushions the growing embryo against knocks. The embryo is joined to the placenta by the umbilical cord, or umbilicus. Blood from the embryo passes down the umbilicus and meets the mother's blood in the placenta. Here, all the food and oxygen needed for the growth of the embryo pass from the mother's to the embryo's blood. At the same time the mother's blood carries away the waste substances from the embryo.

Mammals that bear their young in this way are called placental mammals. All the mammals except the egg-laying monotremes and the pouched marsupials are placentals. Some placentals are born in an advanced state, like a newborn horse, or foal. At the moment of birth, it is about a sixth the size of its mother. It is fully-formed and able to run about before an hour has passed. The only thing it lacks is the ability to feed itself. It is so well developed because it has remained inside its mother's womb, growing and being fed, for nearly a year.

Other mammals are less well developed at birth. They have no hair, their eyes are shut and they are helpless, like rabbits and mice. They stay in a nest and have to be kept warm by their mothers until they have grown and developed.

The baby marsupial needs all this time sheltering and feeding in its mother's pouch because it has hardly started to grow before it is born. It is still an embryo, not fully formed, and little more than a mass of flesh with a sense of smell to guide it to the pouch, and limbs just sufficiently formed to drag itself into the pouch. The newborn baby of a grey kangaroo, for instance, is no bigger than a string bean, although when fully grown the animal is as tall as a man.

Placentals
In all other mammals the baby spends longer inside the mother's body, where it is nourished by the *placenta*. This is a mass of tissue, also called the afterbirth, within which is the embryo. It is called the afterbirth because it comes away from the womb and is shed after the baby

▶A few seconds after its birth, the foal of this Grant's zebra shakes its head and tries to free itself from the protective birth sac. Then it will twist and turn to break the umbilical cord which still attaches it to the mother. After that it will struggle to stand up. Only the young of animals which need to run from predators are active so soon after birth.

THE EMBRYO DEVELOPS

When a sperm enters an egg, as we have seen, the two nuclei join. The egg is then said to be fertilized. This means that growth will take place to produce a new and eventually an entirely independent animal.

Dividing cells
This growth is by repeated division of the egg. First it divides into two daughter-cells. Each daughter-cell starts half as the size of the original egg but it absorbs liquid food through the cell membrane so that it grows until it is full size. This division followed by growth is repeated to produce four, eight, sixteen and thirty-two cells and so on. This is called cleavage and a ball of cells forms which is the beginning of the embryo. At this stage the embryo sinks into the wall of the womb, which has developed to receive it.

Then comes a different stage in the development of the embryo. The cells begin to form groups and each group, by repeated division and growth, begins to form one of the organs of the future animal. The first organ to appear is the spinal cord. Then the future muscles of the back are formed, followed by the legs, tail and head. There is no yolk but the blood vessels are formed at an early stage. They are linked with the mother's blood supply in the placenta and bring food and oxygen to the embryo.

Organs form
After the basic structure of the embryo has formed, it develops all the organs it will have as an adult. At first it looks very different from the adult, but it is very interesting that the embryos of all vertebrate animals (animals with backbones) look the same. At one stage, a human embryo looks like a fish embryo because both have a tail and the beginnings of gills on the head. And a 5-day old chick embryo looks like a 13-day rabbit embryo. Thereafter, it becomes easier to tell the difference between bird and mammal embryos, and then which kind of bird or mammal. In the last stages, the foetus, as it is now usually called, is perfectly formed and is only growing.

Two of the important changes which takes place at birth are the closing of the blood supply through the umbilicus to the placenta and the beginning of breathing through the lungs.

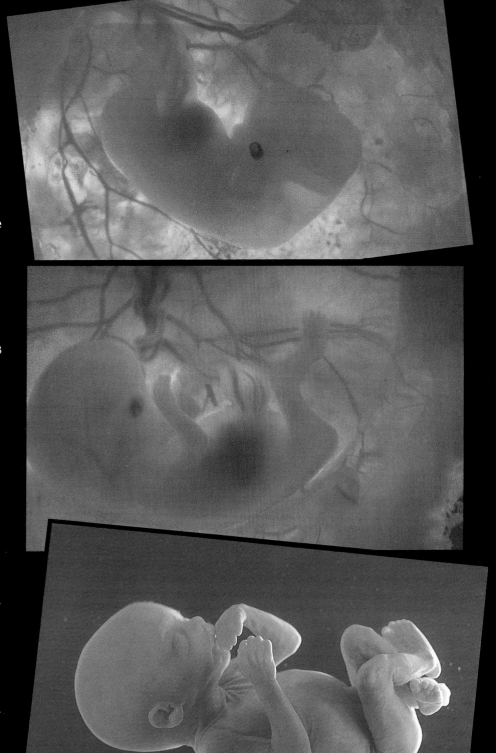

▲The growth of the human embryo: at 7 weeks (**top**), at about 10 weeks (**middle**) and at 12 weeks (**above**).

47

4 GENETICS AND HEREDITY

HEREDITY: VARIATIONS ON A THEME

In any group of animals or plants, no two individuals are the same in every way. All cats are unmistakably cats, and cabbages are easy to recognize as cabbages, but there are slight variations by which we can recognize a particular cat, or, if we were experts, a particular cabbage. Some of the differences, or variations, between individuals are due to their surroundings. For instance, plants which grow in the shade are tall and pale compared with the same kind of plant growing in the full sunlight. Other variations depend on the parents of the individual. When one or both parents of a red-haired child have red hair themselves, we say that the child inherited red hair from its parents. The same is true for the

colour of the eyes, height and so on.

Genetics is the scientific study of variations between individuals and the way these variations are inherited, or passed from parents to offspring. Genetics also helps explain how living things have evolved in so many different forms (page 55).

The discovery of genetics
The way that characteristics are passed from parent to offspring was first discovered by an Austrian monk called Gregor Mendel who lived in the nineteenth century. He grew different varieties of pea and showed that when flowers were pollinated from the same

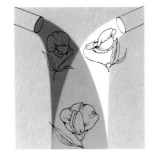

◀ The flower of the pea plant which Gregor Mendel used for his experiments on heredity. He chose the pea plant because it is well protected from foreign pollen, he could breed pure varieties, and it was easy to buy. He studied the colour and form of the seeds and pods, as well as the colour and position of the flowers.

▶Each pea plant receives two instructions about the colour of its flowers – one from each parent. A pure white plant crossed with a pure purple plant gives plants with purple flowers because instructions do not blend (**above**), one always dominates. If these plants are bred together, one out of four offspring has no purple instructions, so its flowers are white.

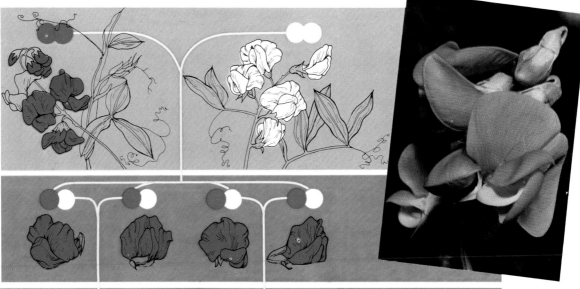

variety, all the next generation of plants were the same as the parents. These are called pure varieties. When two different pure varieties were crossed, the next generation resembled only one parent in colour and height, and so appeared pure. For instance, when a purple flowered variety was crossed with a white variety, all the offspring were purple. However, when this first generation is crossed, the second generation consists of both purple and white peas, but there are three times as many tall plants as dwarfs. The same results are found with other characteristics.

Dominant and recessive characteristics

From such experiments Mendel realized that each animal or plant carried two sets of genetic instructions which determine its characteristics. Each egg or sperm has only one set of the parent's instructions so when they join at fertilization, the zygote has two sets of instructions: one from each parent. This means that exactly half the characteristics of an individual comes from each parent. However only one set of instructions will show. A pea plant with purple and white parents is purple, not pink (see illustration) or white. This is because purple instructions are what scientists call dominant and the white instructions are recessive. And where both dominant and recessive characteristics occur in the same individual it is always the dominant characteristic that develops. In the second generation, some plants have two sets of dominant purple instructions, others have a set of each and are purple, but one quarter of the plants are white because they have two sets of the recessive white instructions. In humans, black hair is dominant over fair hair. If a dark-haired man and a fair-haired woman are pure strains, that is the man has two sets of dominant instructions, their children will be dark-haired.

When animals or plants reproduce asexually, as in budding (page 8) and vegetative reproduction (page 24), the new individuals are exactly the same as their parents. Genetically identical individuals are called *clones*.

HOW GENETICS WORKS

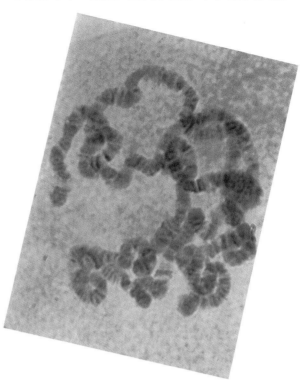

◄The chromosomes of most living things are very small and difficult to see. But in the cells of the salivary glands of the fruit fly *Drosophila* there are giant chromosomes. Because of this, scientists find the fruit fly very useful for the study of genetics. The dark bands across the chromosomes indicate where the genes are found.

Mendel did not know the way genetic instructions worked and they were not discovered until many years later in the nucleus of each cell. Inside the nucleus of each cell in the body there are pairs of long, fine threads called *chromosomes*. Humans have 23 pairs, some plants have over 100 pairs and some flies have no more than eight pairs. The chromosomes carry the genetic instructions. Each set of instructions – for flower height for instance – is called a *gene* and each chromosome carries an enormous number of genes. The gene is made of very complicated molecules called deoxyribonucleic acid (DNA for short). DNA is made of four kinds of smaller molecules which are like letters in a code. The arrangement of the four 'letters' spells out the instructions in the gene.

Mitosis

When a cell divides, during growth of the body or in the binary fission of an amoeba (page 8), each chromosome makes a copy of itself to form an identical pair. There are now two sets of chromosomes in the nucleus and when it divides, one set goes into each half and the two new cells have exactly the same genetic instructions as the original cell. This type of division is called *mitosis*.

Meiosis

When sperm and eggs are made, there is a different form of cell division. The chromosomes double, as before, but the cell then divides twice so that each of the four resulting sperm or eggs have only a single set of chromosomes. This process is called *meiosis*. Later, when a sperm fertilizes an egg, their nuclei fuse and the chromosomes from each pair up (see illustration below right).

►Cell reproduction. By mitosis, cells increase in number.
1 Chromosomes thicken. **2** Each makes a copy of itself so becoming two strands joined at a centromere. **3** Chromosomes collect at the cell centre and a spindle appears at each end or pole. **4** Chromosomes split and **5** the strands move to opposite poles. **6** A cell membrane grows around each group of chromosome strands. **7** The result is two cells identical to the parent cell and with the same number of chromosomes.

Meiosis occurs in the production of sex cells.
1 Chromosomes thicken, **2** become two-stranded and grouped in pairs. **3** Strands in each pair break and rejoin, so mixing genetic material. **4** Pairs separate and move to opposite poles or ends of the cell. **5** The cell divides. **6** Two-stranded chromosomes in each new cell split and the cells divide by mitosis. **7** Four cells result, each with half the number of chromosomes of the parent cell and mixed genetic material. The chromosome number is restored when the sex cells join in sexual reproduction.

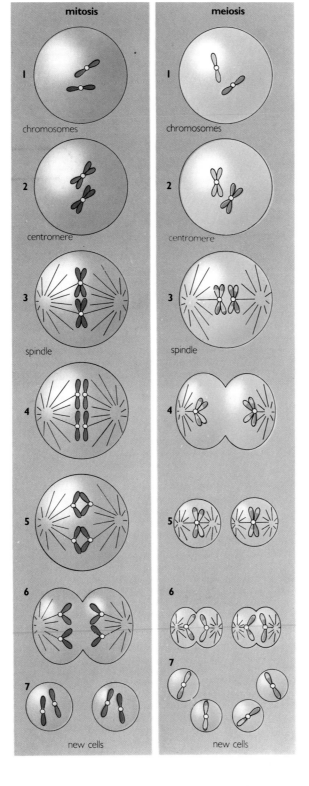

mitosis / meiosis

chromosomes — chromosomes
centromere — centromere
spindle — spindle
new cells — new cells

X and Y chromosomes

In many living things, the sex of the individual depends on one pair of chromosomes. In humans there are two types of chromosome, called X and Y. If a person has two XX chromosomes that person will be female, but if there is one X and one Y, that person will be male. All human eggs have X chromosomes, but human sperms have either X or Y chromosomes, so the sex of a baby depends on which kind of sperm fertilizes the egg.

Mutations

Normally a cell divides to give two exact replicas, with the same genes in each, but sometimes something goes wrong. A change in the genetic content of a cell is called a *mutation*. It generally occurs if the chromosomes are damaged when they pull apart, so that pieces containing some genes are lost or get into the wrong nucleus. This is called a chromosome mutation. A gene mutation is caused by the DNA 'letters' changing and can be caused by X-rays, nuclear radiation or chemicals.

A result of a mutation is that the offspring will be rather different from their parents because the cells have received new instructions from their genes.

▲A true albino is white with pink eyes because it lacks the normal pigment. Albinos, whether tigers (**above**) or blackbirds (**left**) are rare because they are often attacked by their own species and are easy for predators to see.

PRACTICAL GENETICS

There are many different breeds of dog, from tiny chihuahuas to huge mastiffs, curly-haired poodles to short-haired labradors, as well as dogs which are bred for a particular purpose: collies that help shepherds, dobermans for guard duties and hounds for hunting different kinds of animal. All these breeds are descended from one kind of dog and they have been developed, or bred, over a period of hundreds of years. They now look very different from the wolf, which was the wild ancestor of the domestic dog. The same artificial breeding has been used for producing the many kinds of sheep, cattle and poultry, as well as the varieties of crops and garden plants.

When humans first started to keep animals and grow plants, the best were kept for breeding. If, for instance, the peas growing on some plants were found to be larger than those in the rest of the field, they would be kept as seeds for next year's crop. Over the years, the size of peas has increased, so that a crop of peas gives a greater yield. They are now very different from the seeds of wild peas. In the same way a farmer who kept cattle for meat would keep the largest and best bull for mating with his cows.

Hybridization

Improving plants and animals was quite slow until the science of genetics developed. Now much more is known about how qualities are passed from generation to generation, and animals and plants with particular qualities can be produced by the careful selection of their parents. One technique is hybridization. This means that two plants or animals of different varieties are bred together, that is, crossed. The offspring, known as hybrids or crossbreeds, share their parents' characteristics and they are often stronger than either parent – a

▼Crops being grown in plots on an experimental station. Each plot contains one strain, or type, of a crop. The growth of each strain can be measured and compared with that of others growing in the same climatic conditions. The successful strains will be selectively cultivated to improve yields.

▲This chinese lily, *Lilium henryi*, has an open one-colour flower. The hybrid offspring of a cross between these two lilies has the flowers seen in the picture **below**.

▲The trumpet-shaped flower of the species of lily called *Lilium leucanthus*. The brownish petals curl back to show the white inside.

◄A botanical freak – a laburnum (yellow flowers) and a purple broom growing as one tree. It was produced by grafting (asexual reproduction) and not by cross-pollination (sexual reproduction) so the flowers look like each parent and not a mixture of the two.

▲The hybrid offspring of a cross betwen *Lilium leucanthus* and *Lilium henryi* has colourful trumpet-shaped flowers. Many new hybrid varieties are produced by plant breeders. They are often more vigorous than either parent having, for example, more or bigger flowers, or being more hardy.

characteristic called hybrid vigour. Modern types of grain have been produced by hybridization.

Animal and plant breeders try to produce breeds which are easy to keep or grow, or are resistant to diseases as well as ones that produce larger yields of meat, milk, wool, seeds or fruit. Many cows give large quantities of milk when they feed in lush meadows, but poor countries need cows which do well on sparse pastures.

Artificial breeding

Artificial breeding is an important technique. Plant breeders choose which plants they want to cross. Then they collect pollen from one plant with the characteristics they want and place it on the stigma of another.

Animal breeders use artificial insemination. Sperm is taken from a male and placed in the reproductive organs of the female. Because sperm can be stored, one male can fertilize many more females than would be possible naturally. A new technique is to remove the ova from a female, fertilize them in the laboratory and place the embryos in the wombs of other females. This process increases the number of offspring a choice female can produce. It is now used for saving rare species. For instance, the bongo, a rare antelope, is being bred from embyros placed in the wombs of elands, an antelope which is easily kept in zoos.

Genetic engineering

The newest and most important technique is genetic engineering. Instead of selecting the best animals or plants for breeding, genes containing instructions for particular characteristics are transferred from one species to a second so that it acquires the characteristics of the first.

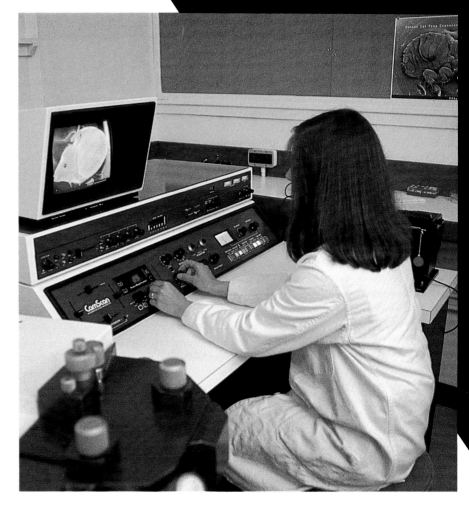

This is becoming a very important method for treating diseases caused by genetic defects. For instance, diabetes is a disease caused by a lack of the hormone insulin which controls the amount of sugar in the blood. Diabetics lack the genes that instruct the body to make insulin and so have to be treated by injections of insulin obtained from animal tissues. This insulin is expensive but cheaper insulin can be made by putting genes for insulin production into bacteria. These bacteria start to make insulin, which is collected and used for the injections.

▲A botanist studies the effects of disease treatment on a plant. Powerful, modern electron microscopes reveal important details of the structure of plants, so helping botanists in their work of treating disease and developing better strains of plant.

▶A mule is a hybrid – the offspring of unlike parents. It is born of a female horse mated with a male donkey. It has the speed of a horse but the hardiness of a donkey.

5 THE HISTORY OF LIFE

So far in this book we have studied the reproduction of plants and animals and genetics. Reproduction shows how the lives of individual living things begin, while the science of genetics explains why individual members of the same species are not identical. But when did the very first living things appear, and how did they reach their present forms?

THE VERY FIRST LIFE

Life on earth started about 3,000 million years ago. No one knows how life began or what the first living things looked like, because the earliest fossils we have are of already well developed animals. Many scientists believe that the first life started in a chemical 'soup'. In some way that we can only guess at, chemicals came together to form blobs of jelly-like substances that could grow and reproduce. Eventually bacteria-like organisms developed and from them simple plants and animals.

Development of living things was very slow until about 600 million years ago. Then, scientists believe, the climate became warmer and there was more oxygen for breathing. Many kinds of animals appeared at this time: sponges, jellyfish, corals, worms and shellfish, which are preserved as fossils in rocks. As time passed new animals and plants came into existence while others disappeared.

THE THEORY OF EVOLUTION

Until about 200 years ago everybody supposed that each species, or kind, of animal and plant, had been separately created and that species were fixed and unchanging. As modern science began to develop, the idea was put forward that each species had arisen from a previous species by a process of gradual change. Scientists called this process 'evolution' from a Latin word meaning to unfold or roll out. Each change is usually very

▼The early family tree of the dogs is not certain. They may be descended from wolves, from coyotes, or from some unknown extinct animal. But their most distant ancestors were probably the small extinct creatures called *Tomarctus* and *Cynodictis*. By breeding dogs of the qualities he values, man has now created over 100 varieties.

old English sheepdog

cocker spaniel

eskimo

pekingese

chihuahua

pug

pomeranian

working collie

alsatian

coyote

jackal

European wolf

Cynodictis

Tomarctus

54

mall so that the evolution of one species from another is very gradual and can be seen only by studying fossils. These show changes which have taken place over millions of years. Only rarely does such a change take place fast enough for us to become aware of it.

The horse is a good example of how one animal has evolved. The first horse, called *Hyracotherium*, lived 60 million years ago. It was the size of a fox terrier, had four toes on the front feet, three on the rear feet, and ate the leaves of trees. *Hyracotherium* died out and its place was taken by several kinds of larger animals which had fewer toes. This sequence continued until finally we had the modern horse, with a single toe ending in a stout nail, the hoof, on each leg. The teeth also evolved and the horse changed from plucking the leaves of trees to cropping tough grasses.

Understanding evolution is important because it shows how organisms are related. Animals and plants are very different from each other but there is reason to believe they evolved from a common ancestor. Similarly, as we know from comparing the structure of their bodies, birds and mammals both evolved from reptiles.

WORKING OUT RELATIONSHIPS

Nobody knows exactly how many different kinds of living thing there are. One and a half million species of animals have been given names but there may be as many as three or four million more. There are 750,000 known species of insects alone, and thousands more are discovered every year. New birds or mammals are found only occasionally. There are not so many species of plants: only about 250,000 are known.

Species
There are so many sorts of living thing that naming them has always been a problem. Three hundred years ago biologists started to classify animals and plants, or put them in order, by grouping them together. Any group of animals or plants which look and behave like each other but differ from all others is called a species. Members of a species breed among themselves but not with other species. The appearance and behaviour of an organism depend on its genes (page 49). All members of the same species have the same pattern of genes, so when they mate the same pattern is passed on to the

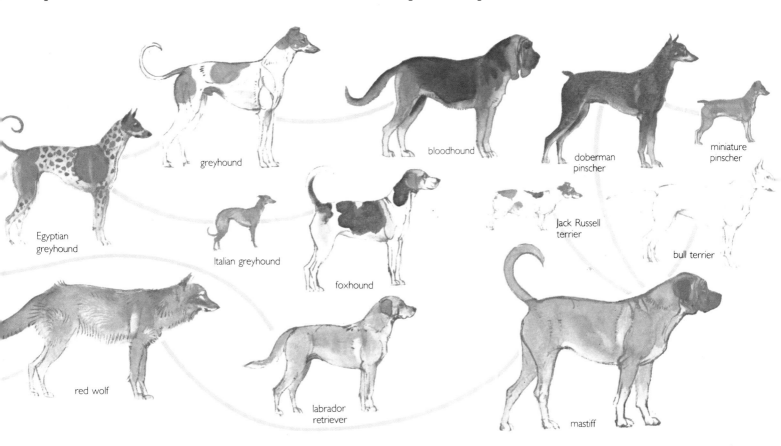

Egyptian greyhound

greyhound

Italian greyhound

foxhound

bloodhound

doberman pinscher

miniature pinscher

Jack Russell terrier

bull terrier

red wolf

labrador retriever

mastiff

next generation, which will look and behave like their parents.

The structure and behaviour of different animals and plants normally prevents them from mating with members of other species. Sometimes this does happen and two species interbreed but the offspring, called a hybrid, is sterile and unable to reproduce. For example, a mule is the sterile offspring of a donkey and a horse.

Names and classifications
A Swedish naturalist called Carl Linné (though he is better known by the Latin form of his name, Carolus Linnaeus), who lived in the eighteenth century, invented the system, which is still used, for giving

animals and plants scientific names. He realized that some species are very similar and he grouped them into units called genera (singular – genus). Each organism has a two-word name, based on Latin or Greek words, and the first name shows the genus. The second word is the specific name and shows which species the organism belongs to.

For instance, the domestic dog is called *Canis familiaris* and the wolf *Canis lupus*. These names show that scientists have worked out that wolves and dogs are quite closely related.

Linnaeus' system of naming is important because it shows how animal and plant species are related to each other. Dogs are descended from wolves, and the genus name *Canis* shows their close relationship. The more closely related two species are the more features they have in common. Foxes and all the other members of the dog family are rather different from dogs and wolves but they all evolved from a common ancestor. Scientists believe that, even further back in time, there was a common ancestor which linked the dog family with the families of cats, bears, badgers and other flesh-eating or carnivorous mammals. Taking this theory to its logical conclusion, scientists believe that all members of the animal kingdom are related to each other, and that, somewhere at the beginning of the earth's history, there was an organism from which both animal and plant kingdoms have descended.

▶In these two diagrams the limb-bones of various vertebrates are compared. (**Right**) The bones of the human arm, of the flipper of a dolphin and the wing of a bird look very different, but the basic plan is the same in each. (**Below**) The bones of the human leg, of the hind-legs of a horse and of a frog also differ markedly yet have the same basic plan. This supports the theory that all these animals, indeed all vertebrates, are descended from a common ancestor.

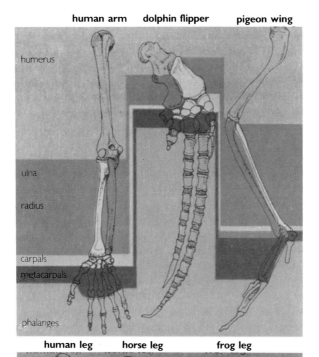

human arm   dolphin flipper   pigeon wing

humerus

ulna

radius

carpals

metacarpals

phalanges

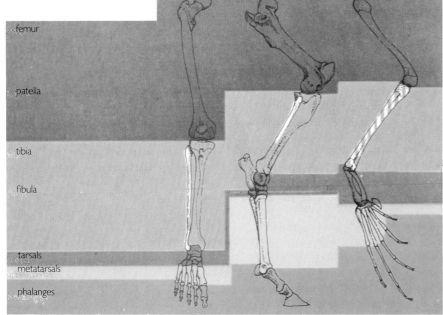

human leg   horse leg   frog leg

femur

patella

tibia

fibula

tarsals
metatarsals

phalanges

EVOLUTION AND NATURAL SELECTION

One species changes into another during the course of evolution, through a process scientists call natural selection. The way scientists think that this works was described by Charles Darwin in his book, *On the Origin of Species by means of Natural Selection*, published in 1859.

Animals and plants produce many offspring, yet their numbers remain roughly the same. Although an oyster produces millions of eggs, an oak tree thousands of acorns and a robin ten or more

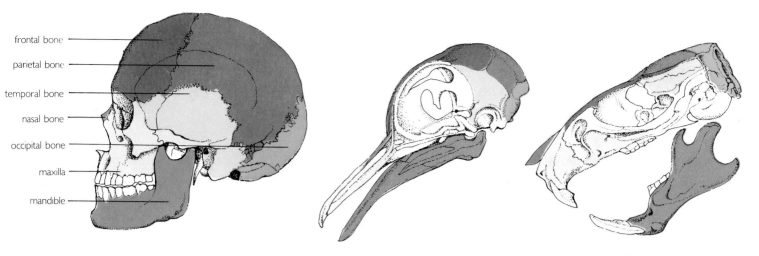

frontal bone
parietal bone
temporal bone
nasal bone
occipital bone
maxilla
mandible

▲A comparison of the skulls of a human (**far left**), a pigeon (**middle**) and a rat (**above**). Each skull has the same bones, but they vary in size and position according to the animal's way of life. For example, a rat's sense of smell is important, so its nasal cavity (nostril) is larger than a human's or a bird's.

ggs each year, most die before they grow up. Therefore there must be a struggle for survival if only sufficient offspring survive to replace their parents and keep the population level. There is competition for food and a place to live, so many starve. Some individuals are eaten, while others are killed by bad weather and accidents.

An aid to survival

Because not all individuals in a species are the same (page 49), some will have better characteristics for helping them to survive. They may be larger and stronger or better at finding food, so they will be more likely to survive and reproduce. Heredity will then ensure that their characteristics are passed to their offspring.

When the environment is steady, species do not change and the process of natural selection weeds out individuals which are unsuitable. Then, when the food supply changes or the climate alters, the species adapts because some individuals are better at surviving the new conditions and so live to breed. Sometimes the change is so great that a new species is formed.

A new form of an animal appears through a mutation (page 50). Most individuals with mutations die because they are not properly adapted to their conditions. But if the conditions change they may be better adapted than the normal individuals which now die out.

EVOLUTION IN ACTION

Scientists believe that evolution is usually very slow and hundreds of generations may pass before any change in a species

can be seen. Very occasionally, however, rapidly-breeding species evolve fast enough for us to see the changes. The development of bacteria which are resistant to drugs is one example. Another is those species which become resistant to poisons. Warfarin is a very powerful poison which is used for getting rid of rats from farms and buildings. But a few rats are immune to the poison. They and their offspring survive and when the other rats have died, these immune rats spread. In some places there are now whole populations of rats which cannot be killed by Warfarin.

In the plant world, rapid selection can be seen in the colonization of spoil tips in mining areas. These huge masses of rock disfigure the countryside but very little will grow on them because they contain high concentrations of lead, copper, arsenic and other poisonous substances. However, it was found that some individual plants are immune to the poisons. When ordinary bent-grass seeds are planted on tips, they will germinate but most die because their roots do not develop in the poisoned soil. Only one in a thousand plants will grow to maturity. Thus tolerant varieties can be developed by selective breeding.

Adaptation is a serious problem for doctors when germs become resistant to drugs. When patients are treated with bacteria-killing, or antibiotic drugs, such as penicillin, some of the bacteria survive. They are not killed by the penicillin and they breed rapidly because there is no competition from other bacteria. So, very soon, there is a new strain of bacteria and doctors have to find new drugs which will attack the bacteria and cure the disease.

The importance of pre-adaptation

Rats, grasses or bacteria which are immune to poison are said to be 'pre-adapted'. They have characteristics which have no use in their normal environment but which help them survive when the environment changes. Pre-adaptations play an important part in evolution. In the ear of a mammal there are three small bones, the auditory ossicles, which transmit sound from the eardrum to the sense organs of the inner ear. One of these ossicles performed this function in reptiles while the other two were part of the jaw hinge. As some reptiles evolved into mammals, the jaw hinge became stronger by displacing these two bones. They were now free to become part of a more efficient ear mechanism.

Sickle-cell anaemia

An example of natural selection in humans is the disease called sickle-cell anaemia, in which the red cells in the blood are misshapen and unable to carry oxygen properly. Sickle-cell anaemia is caused by a recessive gene and makes the sufferer short of breath and, in serious cases, leads to death. However, people with sickle-cell anaemia do not suffer so severely from malaria as normal people, because the protozoan causing the malaria cannot live in the misshapen cells. In West Africa, where malaria is common, 20 per cent of the population has sickle-cell anaemia. In the United States there are many people whose ancestors came from West Africa, but only 9 per cent of these people have the anaemia. But as there is no malaria in the United States, sickle-cell anaemia is a disadvantage.

Tetraploidy

Plants can produce new species more quickly than animals by a process called tetraploidy. Normally individuals of two different species do not interbreed or, if they do, their offspring are sterile, like a mule (page 53). They are sterile because their chromosomes are a mixture from each parent and not properly paired. So they cannot divide into equal sets at meiosis (page 50). They, therefore, fail to form germ-cells or set seed. On some occasions, however, a plant hybrid produces germ-cells which have double the normal number of chromosomes because meiosis has not taken place. If two such germ-cells fuse, their two sets of chromosomes can now form pairs. Then when meiosis takes place, the chromosomes can divide equally, normal germ-cells form and proper fertilization takes place.

A new species of grass was created by tetraploidy on the south coast of England in 1870. The native cord-grass grew on muddy shores and it interbred with a species which came from North America and formed a new species which spread rapidly. The English species had 56 chromosomes in its cells, the North American species had 70 chromosomes. A normal hybrid would not have pairs of chromosomes, but this new species was a tetraploid and had 126 chromosomes, that is a complete set from each parent. These could divide properly at meiosis, so that breeding could take place.

Changing to survive

A famous example of a species changing is the peppered moth. It gets its name from its mottled grey wings which make it very difficult to see when it rests on lichen-covered tree trunks. There is a mutant form of the peppered moth which is black. It first appeared in 1848 and it was extremely conspicuous against naturally coloured trunks. Birds could find it easily and it remained very rare. Then, as the number of factories increased, smoke from their chimneys blackened the trees and killed the lichens in industrial regions. The black peppered moth became hard to see and survived, whereas the normal grey moths were easily caught by birds. Natural selection was favouring the black moths so they became very common and the original form became rare. Now, however, because there are strict controls on air pollution from factories, the original kind is becoming more common.

Developing in isolation

The two kinds of peppered moths are still one species and can interbreed. They could develop into two species if they became separated. Once in isolation, each would undergo its own mutations and be changed by natural selection until the two became so different that they could no longer interbreed if they came together again.

This sort of isolation takes place when two populations are separated by a barrier which they cannot cross. The barrier could be a moutain range, a desert, the sea or open country between forests. Animals living on islands are often different from relatives on the mainland. For instance, animals living in Britain are frequently different from those living in the rest of Europe. The differences are slight because they have been separated for only about 10,000 years, since the English Channel was formed. This is a very short time in evolutionary terms.

The Galapagos Islands
The formation of species in isolation for a very long time is shown by the animals living on the Galapagos Islands in the Pacific Ocean. The Galapagos animals are different from species living in South America, 600 miles to the east. The islands are the home of the flightless cormorant and the marine iguana which are found nowhere else, but some of the islands also have their own unique species. For instance, there are 14 different types of Galapagos tortoise. Some live only on one island but five live on the largest island. However, each of the five lives on a different volcano and, because the tortoise cannot travel from one volcano to the next, the volcanoes are a kind of island which isolate the tortoises.

Adaptive radiation
The 13 species of Galapagos finches have evolved from a single species which arrived from the mainland centuries ago.

▼Cormorants are found all over the world. All can fly except the flightless cormorant, with very weak wings, which is found on two of the Galapagos Islands in the Pacific.

◄Another of the animals unique to the Galapagos Islands is the giant tortoise. Because they live on different islands or in isolated areas, 14 different types have evolved.

In the isolation of the islands, new species have evolved with different feeding habits. Some eat seeds, others eat insects and the woodpecker finch uses twigs to pick insects out of holes and crevices. The evolution of many species with different habits from a single ancestor is called *adaptive radiation*. The adaptive radiation of the finches on the Galapagos Islands took place because, when the ancestral finch arrived, there were plenty of ways of life, such as food and nesting sites, unused by other birds. Its descendants could therefore evolve adaptations to fit these ways of life.

Much of the evolution of the animal and plant kingdoms has been through adaptive radiation. For instance, when reptiles arose from amphibians by developing a watertight skin and a watertight eggshell, they could invade the land and occupy many ways of life which the amphibians were unable to exploit.

Sometimes an enormous amount of adaptive radiation takes place in a small area. In Lake Victoria there are about 200 species of cichlid fish, half of which are found nowhere else. They differ mainly in their feeding habits. There are species which feed either on plants, crustaceans, snails, water insects or the larvae of other fish. One species even scrapes the scales off other fish. This diversity of species has evolved over the last 750,000 years during periods when the water level dropped and Lake Victoria was split into many small isolated lakes.

Parallel and convergent evolution

Occasionally two groups of animals that are closely related evolve along similar lines. For instance, the marsupial, or pouched, mammals of Australia are isolated from the placental mammals living in the rest of the world. As a result the two groups of mammals have evolved species with similar adaptations for similar ways of life. The thylacine or Tasmanian wolf is a hunting animal similar to a true wolf. The kangaroos and wallabies are grass-eaters like the placental deer and antelopes. There are also marsupial equivalents of rats, mice, rabbits and moles. This is parallel evolution.

Convergent evolution is the process by which unrelated species evolve to resemble each other by taking up similar ways of life. The prehistoric ichthyosaurs were reptiles which lived in the sea. They looked and behaved like present-day dolphins, which are mammals. Both kinds of animal are similar to fish because they have flippers like a fish's fins and swim by beating their tails.

Co-evolution

Many animal species are dependent on particular plants for food or shelter: the giant panda cannot exist without bamboo shoots to eat. Some plants are dependent on animals: when clover was introduced to Australia bumblebees had to be brought in as well to pollinate the flowers. The fossil record shows that many flowering plants evolved at the same time as certain insect groups. These plants developed flowers which were pollinated by insects rather than by the wind. The flowers had colours to which insect eyes are sensitive and they secreted nectar as a 'reward' for insects which flew from flower to flower transferring pollen. Flowers of mints and orchids have a platform for bees to land on and the pollen is at the bottom of a long tube so that the bees have to force their way past the stamens and stigma. At the same time as the flowers were evolving, the insects were developing specialized mouthparts, such as the hollow tongues of butterflies and moths, for collecting nectar and pollen.

Some flowers attract a wide range of insects but others are available only to certain insects. Only bumblebees can force open snapdragon flowers and honeysuckles open at night to attract moths. There is an amazingly close relationship between orchids, like the bee and fly orchids, which resemble female insects, and even smell like them. Male insects land on the flower and try to mate with it, and so pick up or deposit pollen. In relationships like these both species, plant and insect, depend completely upon each other. If either changes in any way it could be disastrous for both. As a result, such relationships are generally the exception rather than the rule.

▶Orchids are treasured flowers because they are often rare as well as beautiful. They are also sought after because of the strange shapes of their flowers. This fly orchid is pollinated by one species of fly. The orchid's flowers look like the female fly. The male fly tries to mate with it and so carries away pollen to the next fly orchid it visits.

61

GLOSSARY

Adaptive radiation The development of many species with different habitats from a single ancestor.

Alternation of generations A two-part life cycle found among some plants when a sexual stage is followed by an asexual stage.

Anther The part of the stamen of a plant containing the pollen.

Binary fission Form of asexual reproduction in which one cell divides into two parts.

Carbohydrate Energy-giving substance formed from the gases carbon, hydrogen and oxygen. Carbohydrates are found in starch, sugar and glucose.

Carpel The female parts of a flowering plant: the stigma, style and ovary.

Chromosome Microscopic thread-like structure inside the nucleus of all kinds of living cell.

Chrysalis The pupal stage in the development of beetles, moths and butterflies.

Cilia Short hair-like structures that stick out from many kinds of living cell.

Clones Individual plants or animals that have identical genes.

Cross-pollination When pollen from one flower is carried to another flower of the same type and fertilizes (pollinates) it.

Embryo The individual living thing that develops from a fertilized egg cell.

Flagellum A long hair-like structure that sticks out from the cell of some minute organisms. The beating of the flagellum helps the creature to move through the water in which it lives.

Foetus The unborn or unhatched offspring of animals. A foetus is the next stage in development after the embryo and in humans is used to describe embyros more than eight weeks old.

Fruit The part of a plant that carries the seed or seeds. There are several different kinds of fruit. A true fruit develops from the ovary of the flower. There are two kinds of true fruit: a drupe, like a cherry, which has a single seed (the stone or pit), and a berry, like a cranberry or blueberry that has a cluster of seeds inside one fleshy covering. A blackberry is a collection of drupes clustered round a central core. Each drupe has its own seed. A false fruit develops from parts of the flower that are not the ovary. A rose hip is an example. The flesh forms from tissue around the ovary.

Gamete One of the special cells that, in sexual reproduction, joins with another gamete to form the single cell, or zygote, from which the new individual plant or animal will grow.

Genes Complicated chemical substances contained in chromosomes. Each gene contains instructions about how the cell which contains it should develop.

Invertebrate An animal without a backbone.

Larva (plural **larvae**) The young of animals such as insects and crabs. In general, larvae do not look like the adult creatures into which they will grow.

Marsupials Mammals, the females of which have a pouch in which the babies develop after they have been born and in which they stay until they are old enough to fend for themselves.

Meiosis Type of cell division that occurs when eggs and sperm are made. Each cell divides twice, so each of the four resulting cells only has a single set of chromosomes. Then, when fertilization occurs and the nuclei of the sperm and egg fuse (join), the new cell has a double pair of chromosomes again.

Membrane Thin tissue surrounding individual cells. It is not always solid, and substances such as chemicals can pass through it.

Metamorphosis A complete change in the shape and appearance of a creature during the course of its development, for cxample when a caterpillar turns into a pupa.

Mitosis Type of cell division in which the chromosomes within the nucleus split into two identical parts. Then the nucleus and the cell divide in two, so forming a pair of identical cells.

Monotremes Mammals, the females of

which lay eggs, although the young when they hatch are fed on the mother's milk. Echidnas are monotremes.

Mutation A change in the genetic make-up of a cell.

Nectaries The organs that make a plant's nectar.

Nidicolous Those birds with young that stay in the nest for days or weeks after they hatch.

Nidifugous Birds whose young can leave the nest and run about soon after hatching.

Nucleus The part of a living cell that contains the chromosomes.

Ovoviviparity Producing young from eggs hatched within the body. The eggs hatch just before they are laid, so that the young are born alive.

Ovule The immature fruit of a plant, which will develop into the seed for a new plant when it is fertilized.

Photosynthesis The process by which green plants use the energy from sunlight to build up complex substances from the gas carbon dioxide and water to provide food for themselves.

Placenta The organ within the womb of mammals from which the developing embryo draws its nourishment until it is born. Mammals, the females of which have placentas, are described as placentals.

Prothallus The tiny, often heart-shaped, sexual stage of those plants, such as ferns, that have a two-stage life cycle.

Protoplasm The liquid or jelly-like substance that is the main part of living cells.

Protozoa Microscopic single-celled organisms.

Pupa (plural **pupae**) The third stage in the life of many insects that comes between the larva and the adult.

Self-pollination When the pollen of a plant fertilizes the stigma of the same plant.

Sepals The outer casing of a flower bud.

Species A group of plants or animals that look alike and behave alike and can breed and produce fertile offspring.

Spores Microscopic reproductive cells of some of the simpler plants.

Stamen The male, pollen-bearing, part of a flower.

◀A warbler brings food to its nestlings. Compare these nidicolous birds (page 43) with the young ostriches on page 41.

Stigma The top part of the style.

Style The stalk connecting the stigma and the ovary.

Vegetative reproduction Type of asexual reproduction in some plants. Bulbs, corms and rhizomes are examples of vegetative reproduction in plants.

Vertebrate Animal that has a backbone.

Zygote Cell formed when two gametes fuse (join). The new individual plant or animal will develop from the zygote.

▼A young deer sucks milk from its mother. All adult female mammals have mammary glands which produce milk to feed their newborn young. When it is mature enough the young mammal will look for its own food. Only mammals provide for their young in this way.

INDEX

Acknowledgements

Heather Angel, Austral
High Commission, A–Z
Botanical Collection,
M. Bavestrelli, Biophot
I. Bucciarelli, Camera
Press, P. Castrano, Bru
Coleman, William Cond
Eric Crichton Photos,
D.M.T. Ettlinger, Bob
Gibbons Photography, a
De Graaf/Dregon Bulb
Farm, Dennis Green,
Brain Hawkes, P.A.
Hinchliffe, Holtstudios,
Archivio Igda, Jacana,
Klebsiella, Roger Kohn
A. Margiocco, G. Mazza
Nature Photographers
N.H.P.A., Oxford Scien
Films, Sandra Pond,
Premaphotos Wildlife,
Richard Revels, John
Robinson,
F. Sauer, Science Phot
Library, Seaphot, Harr
Smith Collection, John
Walsh, Trever Wood,
Zoological Society of
London.